METHODS in NEUROENDOCRINOLOGY

CRC Press
METHODS IN THE LIFE SCIENCES

Gerald D. Fasman - Advisory Editor
Brandeis University

Series Overview

Methods in Biochemistry
John Hershey
Department of Biological Chemistry
University of California

Cellular and Molecular Neuropharmacology
Joan M. Lakoski
Department of Pharmacology
Penn State University

Research Methods for Inbred Laboratory Mice
John P. Sundberg
The Jackson Laboratory
Bar Harbor, Maine

Methods in Neuroscience
Sidney A. Simon
Department of Neurobiology
Duke University

Joseph M. Corless
Department of Cell Biology,
Neurobiology and Ophthalmology
Duke University

Methods in Pharmacology
John H. McNeill
Professor and Dean
Faculty of Pharmaceutical Science
The University of British Columbia

Methods in Signal Transduction
Joseph Eichberg, Jr.
Department of Biochemical and Biophysical Sciences
University of Houston

Methods in Toxicology
Edward J. Massaro
Senior Research Scientist
National Health and Environmental Effects Research Laboratory
Research Triangle Park, North Carolina

CRC Press
CELLULAR AND MOLECULAR
NEUROPHARMACOLOGY

Joan M. Lakoski, Advisory Editor

The CRC Press *Cellular and Molecular Neuropharmacology Series* provides the reader with state-of-the-art research methods that address the cellular and molecular mechanisms of the neuropharmacology of brain function in a clear and concise format. Topics covering all aspects of neuropharmacology are being reviewed for publication.

Published Titles

Molecular Regulation of Arousal States, Ralph Lydic

Neuropharmacology Methods in Epilepsy Research, Steven L. Peterson and Timothy E. Albertson

Methods in Neuroendocrinology, Louis D. Van de Kar

METHODS in NEUROENDOCRINOLOGY

Edited by

Louis D. Van de Kar

Department of Pharmacology and Experimental Therapeutics
Loyola University of Chicago
Stritch School of Medicine
Maywood, Illinois

CRC Press

Boca Raton Boston London New York Washington, D.C.

Library of Congress Cataloging-in-Publication Data

Methods in neuroendocrinology / [edited by] Louis D. Van de Kar.
   p. cm. -- (Cellular and molecular neuropharmacology)
   Includes bibliographical references and index.
   ISBN 0-8493-3363-6 (alk. paper)
   1. Neuroendocrinology--Methodology. I. Van de Kar, Louis D.
II. Series: CRC Press methods in the life sciences. Cellular and
molecular neuropharmacology.
   [DNLM: 1. Neurosecretory Systems--physiology. 2. Neurosecretion-
-physiology. 3. Endocrine Glands--physiology. 4. Hormones-
-physiology. 5. Neuroendocrinology--methods. WL 102 M5934 1998]
QP356.4.M48 1998
612.8--dc21
DNLM/DLC
for Library of Congress   98-3241
                              CIP

Preface

Neuroendocrinology is a scientific discipline concerned with the inter-relationship between neuronal and endocrine systems. Thus, neuroendocrine research addresses the regulation of hormone secretion by the brain as well as the influence of hormones on brain function. Because neuroendocrinology is transitional between endocrinology (the study of hormones) and neuroscience (the study of neuronal systems), it often is lost between these two disciplines. Many classical endocrinologists view neuroendocrinology as neuroscience while many neuroscientists view neuroendocrinology as endocrinology.

This book is intended for students, pre- and post-doctoral fellows, medical residents and others who are interested in neuroendocrine research. The presentation of the chapters is designed to allow the readers to obtain insight into the many scientific disciplines that comprise neuroendocrine research. Thus, the chapters in the book will describe experimental approaches that bridge endocrine and neuroscientific research.

To help in reading this book, a few basic definitions may help guide the reader. *A hormone* is a substance that is released from one organ into the blood stream, it travels in the blood and then induces a physiological reaction in another organ. *An endocrine gland* is the organ that secretes the hormone. *Feedback* refers to a sequence of events that begins with the secretion of the hormone and results in the inhibition (or stimulation) of further secretion of the same hormone.

The hypothalamus is the part of the brain most intimately involved in the regulation of hormone secretion. It is located at the base of the brain on the midline, ventral to the thalamus and dorsal to the pituitary gland. The hypothalamus regulates the pituitary gland. The pituitary gland, in turn, controls the function of several endocrine glands, such as the adrenal gland, thyroid gland, testes and ovaries. The pituitary gland is divided into three lobes: the anterior, posterior and intermediate lobes. The hypothalamus controls the secretion of hormones from the distinct lobes of the pituitary gland through different mechanisms as detailed below. The anterior lobe of the pituitary gland secretes the following hormones: luteinizing hormone (LH), follicle stimulating hormone (FSH), adrenal corticotropic hormone (ACTH), thyroid stimulating hormone (TSH), growth hormone (GH) and prolactin. Hypothalamic releasing factors that are released from the hypothalamus into the pituitary portal blood vessels reach the anterior lobe of the pituitary gland and stimulate the secretion of ACTH, FSH, GH, LH, TSH and prolactin. In addition, the secretion of some hormones, notably prolactin and GH, is kept under tonic inhibition by hypothalamic inhibiting factors that reach the pituitary via the pituitary portal vessels. Hormones secreted by the anterior lobe of the pituitary gland can also act as trophic hormones that regulate the growth and overall function of the target glands.

The posterior lobe of the pituitary gland is also known as the neural lobe because it is an extension of neurons originating in the hypothalamus. Hypothalamic cells containing vasopressin and oxytocin send their axons into the posterior pituitary where they release these hormones from their nerve terminals into the circulation. The intermediate lobe of the pituitary gland contains high concentrations of various peptides, including endorphin, melanocyte stimulating hormone (MSH) and ACTH but their function in humans (if any) is not yet known.

Many brain regions that are involved in regulating emotions (i.e., the limbic system) and homeostasis have afferent and efferent connections with the hypothalamus. These regions include the amygdala, hippocampus, bed nucleus of the stria terminalis, dorsal raphe nucleus, locus coeruleus, parabrachial nucleus and nucleus of the solitary tract. Neurons located in these brain regions can convey information relevant to neuroendocrine control to hypothalamic neurons and receive information from hypothalamic neurons. The processing and integration of neuronal information flow relevant to the secretion of hormones is a primary focus of neuroendocrine research.

A major function of the neuroendocrine system is to ensure the survival of the individual and the species. Situations that endanger individual survival can be external in origin, for example, pursuit by a predator or the sudden appearance of the boss in the office, or conditions due to endogenous changes, for example the chest pain associated with sudden coronary vascular constriction. Other occurrences, such as famine, might not present an immediate danger to the individual but may endanger the survival of the species. Survival of the individual is regulated by different neuroendocrine response mechanisms than the survival of the species. Hormones responsible for the survival of the *individual* are often referred to as "stress hormones", such as the hypothalamic-pituitary-adrenal axis hormones (ACTH, corticosterone/cortisol), the renin-angiotensin system and epinephrine. The survival of the *species* is more dependent on hormones that regulate reproduction, such as LH and FSH. A complex network of afferent inputs from sensory organs, from the viscera and from other regions in the body convey messages to the brain regarding changes in the external and/or internal environment. The brain integrates these inputs and organizes the appropriate endocrine, behavioral and autonomic responses to these environmental or internal challenges.

This book will delineate some of the many different disciplines that are involved in investigations of neuroendocrine control mechanisms and will indicate what can be learned from different approaches to neuroendocrine research. The chapters in the book also will outline the limits of interpretation for several approaches to neuroendocrine research. Beginning with the most elementary approaches to studying neuroendocrine regulation, chapters in this book will gradually increase in levels of complexity. Chapter 1 will deal with cellular and molecular biological methods in immortalized cell cultures of neurons. Chapters 2 to 4 will present molecular biological and immunocytochemical methodologies in neuroendocrine research. The following chapters (5 to 7) will describe more complex neural circuits using *in vitro* slice preparations. Chapters 8 to 10 will explain how to study neuroendocrine control mechanisms and the influence of stress in conscious experimental animals. Analyses of the pulsatile nature of hormone release, both in experimental animals and in humans, will be described in Chapter 11. The final chapter will describe the investigation of neuroendocrine function in humans and how hormone responses to specific stimuli can be used to obtain information about the functional status of the brain, for example in patients who suffer from depression and schizophrenia.

It is my hope that the chapters in this book will encourage readers who are not involved in neuroendocrine research to pursue this exciting area. For those among the readers who are performing research in neuroendocrinology, the following chapters might provide insight into the value of methodological approaches that are not currently employed in their laboratories.

Louis D. Van de Kar

The Editor

Louis D. Van de Kar, Ph.D. is a Professor of Pharmacology in the Department of Pharmacology and Experimental Therapeutics at Loyola University of Chicago, Stritch School of Medicine, in Maywood, Illinois.

Dr. Van de Kar received his B.S. in Chemistry from the University of Amsterdam in The Netherlands, in 1971. He obtained his M.S. in Biochemistry, from the University of Amsterdam (The Netherlands) in 1974. Subsequent to his M.S. degree, Dr. Van de Kar joined the Netherlands Central Institute for Brain Research in Amsterdam where he started his research in neuroendocrinology under the direction of Professor J. Ariens Kappers. In 1975 he was awarded the Fullbright-Hays Travel Fellowship and became a graduate student in the Ph.D. program at the University of Iowa. He obtained his Ph.D. in pharmacology in 1978 under the direction of Dr. Lucas S. Van Orden III. Dr. Van de Kar subsequently joined the laboratory of Dr. William F. Ganong at the University of California at San Francisco, where he was a post-doctoral fellow until 1981.

At the end of 1981, Dr. Van de Kar joined the Department of Pharmacology and Experimental Therapeutics at Loyola University of Chicago, Stritch School of Medicine, as an assistant professor. He received tenure and promotion to associate professor in 1987, and became full professor in 1992.

Dr. Van de Kar is a member of the American Society of Pharmacology and Experimental Therapeutics, American Physiological Society, Society of Neuroscience, International Society for Neuroendocrinology and The Serotonin Club. He is a member of the council of the Serotonin Club and a member of the CNS Section Steering Committee of the American Physiological Society and the Liaison With Industry Committee of the American Physiological Society.

Dr. Van de Kar has delivered many lectures in invited seminars and international symposia, and was a Grass Foundation Traveling Scientist Program lecturer. He also orgainized and chaired several symposia under the auspices of organizations such as FASEB, The Serotonin Club, the International Society for Psychoneuroendocrinology, and the World Congress of Biological Psychiatry.

Dr. Van de Kar is a recipient of research grants from the National Institutes of Health and the United States-Israel Bi-national Science Foundation. He has published over 90 papers in peer-reviewed journals, several book chapters, and over 100 abstracts. His current research interests focus on neuroendocrine aspects of scrotonin receptors and their signal transduction mechanisms.

Contributors

György Bagdy
National Institute of Psychiatry
  and Neurology
Budapest, Hungary

Béla G.J. Bohus
Department of Animal Physiology
University of Groningen
Haren, Netherlands

Gwendolyn V. Childs
Departments of Anatomy
  and Neurosciences
University of Texas Medical Branch
Galveston, TX

Philip J. Cowen
Wareford Hospital
Oxford, England

F. Edward Dudek
Departments of Anatomy and
  Neurobiology
Colorado State University
Fort Collins, CO

Robert J. Handa
Department of Cell Biology,
  Neurobiology, and Anatomy
Loyola University Chicago
School of Medicine
Maywood, IL

Joan M. Lakoski
Departments of Pharmacology
  and Anesthesia
College of Medicine
Pennsylvania State University
Hershey, PA

Ping Li
Departments of Physiology
  and Pharmacology
Bowman Gray School of Medicine
Wake Forest University
Winston-Salem, NC

Zsolt Liposits
Department of Anatomy
Albert Szent-Gyorgi University
Szeged, Hungary

Istvàn Merchenthaler
Women's Health Research Institute
Wyeth-Ayerst Research
Radnor, PA

Mariana Morris
Departments of Physiology
  and Pharmacology
Bowman Gray School of Medicine
Wake Forest University
Winston-Salem, NC

Richard H. Price, Jr.
Department of Cell Biology,
  Neurobiology, and Anatomy
Loyola University Medical Center
Maywood, IL

Celia D. Sladek
Department of Physiology
Finch University of Health Sciences
Chicago Medical School
North Chicago, IL

Bret N. Smith
Departments of Anatomy
  and Neurobiology
Colorado State University
Fort Collins, CO

Jane E. Smith
Departments of Pharmacology
  and Anesthesia
College of Medicine
Hershey, PA

Janice H. Urban
Departments of Neurobiology
  and Physiology
Northwestern University
Evanston, IL

Louis D. Van de Kar
Department of Pharmacology
Loyola University
Stritch School of Medicine
Maywood, IL

Johannes D. Veldhuis
Department of Internal Medicine
University of Virginia Health
  Sciences Center
Charlottesville, VA

William C. Wetsel
Psychiatry and Behavioral
  Sciences
Duke University Medical Center
Durham, NC

Melinda E. Wilson
Loyola University
Department of Anatomy
Maywood, IL

Contents

Chapter 1

Functional Morphology of the Immortalized Hypothalamic LHRH Neurons

William C. Wetsel, Istvàn Merchenthaler, and
Zsolt Liposits

Contents

I. Introduction

More than thirty-five years ago, McCann and colleagues[1] reported that extracts from rat hypothalamus stimulated the release of luteinizing hormone (LH) from the anterior pituitary. The subsequent isolation, purification and sequencing of this factor revealed that it was a decapeptide that could stimulate secretion of both LH and follicle-stimulating hormone from the pituitary.[2] Due to these properties, this peptide

1

was termed LH-releasing hormone (LHRH) or gonadotropin-releasing hormone. Since its discovery, the morphology, physiology and pharmacology of LHRH neurons have been well characterized. Despite this information, the biochemical, molecular and cellular mechanisms which underlie dynamic aspects related to LHRH morphology and physiology are poorly understood. This paucity of information can be attributed primarily to the constraints imposed by the topology of the LHRH neuronal system and to the low numbers of neurons present in rat brain. LHRH neurons are perhaps the only hypophysiotrophic neurons that originate outside of the central nervous system. These neurons arise from the olfactory placode and they migrate into the septal-preoptic-hypothalamic region of the brain.[3,4] From these locations, axons are sent to the median eminence. Due to these attributes, the LHRH neuronal cell bodies are not confined to a discrete nucleus, rather they are scattered along the base of the brain through several brain regions.[5,6] In addition, the numbers of LHRH neurons in rat brain are relatively low,[6] and only 60 to 70% of these neurons project to the median eminence.[7,8] Hence, it may be the case that LHRH neurons *in toto* are composed of a heterogeneous population of cells that are differentially responsive to stimuli and that may subserve functions beside that of regulating gonadotropin secretion.

Presently, there are two different approaches that lend themselves to the molecular and cellular analyses of LHRH neuronal function. First, Wray and colleagues[9,10] have developed hypothalamic slice-explant cultures as model systems to study regulation of gene expression and neuronal activity. In these cultures, cells can be maintained in an organotypic fashion for long period of time *in vitro*. Importantly, the entire extent of individual neurons (viz., perikaryon, axon, and terminals) can be maintained and, for the most part, the interactions among cells are preserved. This procedure holds considerable promise for studying interactions among cells, especially during cell migration and axon extension. This approach is amenable to immunocytochemical, *in situ* hybridization, electrophysiological, and vital dye studies. Unfortunately, this system is not robust enough to support detailed biochemical and molecular analyses. These investigations can be pursued by a second approach, the immortalized LHRH neuronal cell lines.[11,12] These hypothalamic neurons are homogeneous cell lines that can be propagated and maintained in culture. Since large quantities of these cells can be cultured, these cell lines may serve as an important tool in morphological, biochemical, and functional studies. Moreover, since these cell lines are of neuronal origin, they may also serve an important role in basic neurobiological studies.

II. Genetic Targeting of Tumorigenesis to LHRH Neurons

The feasibility of obtaining immortalized LHRH neuronal cell lines became apparent following two major developments in transgenic mouse technology. First, Mason and colleagues[13] had successfully targeted LHRH neurons for gene therapy in hypogonadal (*hpg*) mice. These mice carry a mutation in the LHRH gene where the distal half of the gene is deleted.[14] Rescue of the phenotype resulted in clear visualization of LHRH and gonadotropin-releasing hormone associated peptide (GAP) immunoreactivities in the hypothalami of *hpg* transgenic mice. Importantly, this therapy was found to be capable of restoring reproduction in these mice.

A second development relates to the ability to target the expression of the simian virus-40 (SV-40) T-antigen oncogene to a given cell such that it becomes immortalized. Under these circumstances, a hybrid gene is constructed that contains the promoter region for a gene of interest and this promoter is fused to the coding region for the SV-40 large-T antigen. The hybrid gene is injected into one-cell embryos and these embryos are implanted into the uteri of pseudopregnant mice. Upon birth, the transgenic mice are closely monitored for the presence of tumors. Using this procedure, Hanahan[15] successfully produced mice that developed pancreatic β-cell tumors. Homogeneous immortalized β-cell lines were derived from these tumors. Recently, a similar procedure has been used to immortalize anterior pituitary cells of the gonadotrope lineage.[16,17]

The successful targeting of a given cell for tumorigenesis requires that the hybrid gene contain two things; a promoter and the coding region for an oncogenic protein. In the cases described above, the promoter of the hybrid gene determines the cell-specific expression and the SV-40 T antigen protein immortalizes this cell. Thus, any transcription factor that activates the endogenous gene will also activate the hybrid. Through this procedure, a cell which expresses the gene of interest (e.g., LHRH) will become immortalized through co-activation of the hybrid gene.

Both Mellon and colleagues[12] and Radovick and co-workers[11] constructed a hybrid gene containing a portion of the rat or human promoters and the coding region for the SV-40 T-antigen. Following the injection of this hybrid gene into one-cell embryos, Mellon et al.[12] obtained eight transgenic mice (7 female, 1 male) (Figure 1). All of the mice had underdeveloped gonads and accessory sex organs, and they were infertile. Two mice (females) developed anterior hypothalamic tumors. The tumor from one of these mice was excised, the cells were dispersed and, through a series of differential plating procedures, a homogeneous population of cells was obtained. These GT-1 cells were subcloned by serial dilution to produce three different subclones, the GT1-1, GT1-3, and GT1-7 cells. Upon examination of these cells under the microscope, the GT1-7 cells appear to look the most like neurons, the GT1-3 cells are round in shape but may contain some short processes, and the GT1-1 cells appear to be intermediate in shape between the other two subclones. When these cells are grown on matrigel-coated plates or maintained under low serum conditions, all three cell lines readily adopt a "neuronal" appearance. In particular, cells from all three subclones are visually similar to LHRH neurons *in vivo*.[12,18,19] In my experience over the years, all three cell lines contain transcripts for essentially the same genes; however, the levels of expression may vary among the cell lines.

In comparison to Mellon and colleagues,[12] Radovick and co-workers[11] constructed a hybrid gene that used the human promoter for LHRH (Figure 2). Three transgenic mice were produced; two mice (1 male, 1 female) were infertile, while a third (female) was successfully mated. From two matings, seven of the nine

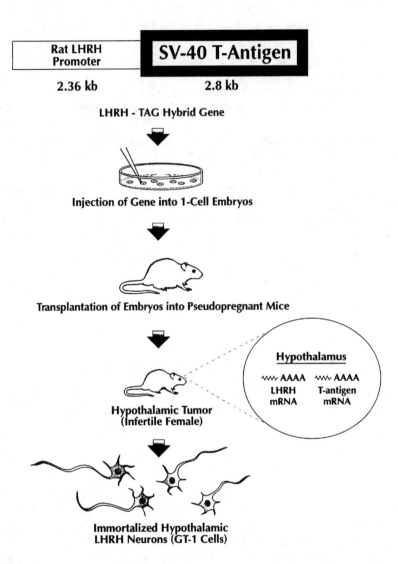

FIGURE 1

The strategy for genetic targeting of LHRH neurons for tumorigenesis using the rat LHRH promoter. Mellon et al.[12] constructed a hybrid gene containing the 5' flanking region of the rat LHRH promoter (−2987 to −1172 bp appended to −441 to +104 bp) that was fused to the entire coding region for the SV-40 large and small T-antigens. This gene was injected into one-cell embryos which were transplanted into pseudopregnant mice. The mice developed anterior hypothalamic tumors. A tumor from one of these mice (female) was excised, the cells dispersed and purified to produce the GT-1 cell line. These GT-1 cells were subcloned by serial dilution to yield GT1-1, GT1-3 and GT1-7 cells. All of these cells immunostain for LHRH. (Adapted from Wetsel, W. C., *Cell. Mol. Neurobiol.,* 15, 43, 1995. With permission.)

T-antigen positive mice were infertile. Four of these mice developed tumors in the olfactory bulb and in all cases, the preoptic-hypothalamic region was devoid of LHRH neurons. An olfactory bulb tumor from one of the F1 progeny (male) was excised, the cells were dispersed, purified, and two cell lines (GN and NLT) were

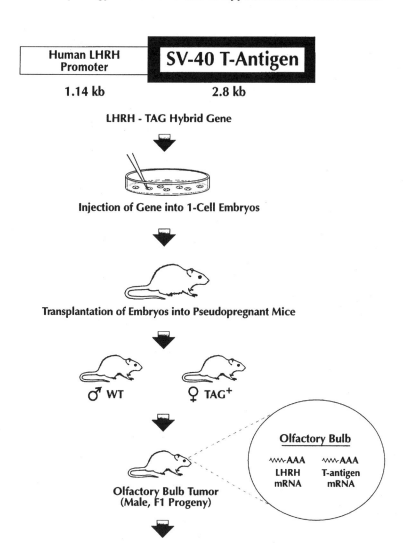

FIGURE 2

A strategy for genetic targeting of LHRH cells for tumorigenesis using the human promoter. Radovick et al.[11] injected a hybrid gene containing the 5' flanking region (–1131 to +5) from the human LHRH gene that was linked to the coding region for the SV-40 T-antigen. One of the females carrying the T-antigen transgene was mated. A male offspring from the F1 progeny developed a tumor in the olfactory bulb. This tumor was removed, and the cells were dispersed and purified to produce the GN and NLT cell lines. These cells were immunopositive for LHRH. (Adapted from Wetsel, W. C., *Cell. Mol. Neurobiol.*, 15, 43, 1995. With permission.)

obtained. All cells immunostain for LHRH, are round or spherical in appearance, and are somewhat larger in size than LHRH neurons *in situ.*[11] While both GN and NLT cells release LHRH into the medium, the NLT cells secrete at least 10- to 20-fold more LHRH than the GN cells.

Since developing the immortalized LHRH cell lines, both Mellon's and Radovick's lab have taken two different approaches to locate the region of the LHRH promoter that is responsible for cell-specific expression. Using the GT1 cells and deletion-reporter constructs, the neural specific enhancer of the rat LHRH promoter has been identified.[20] Importantly, this region of the promoter is capable of targeting expression to hypothalamic LHRH neurons.[21] By comparison, the region of the human LHRH promoter responsible for conferring cell-specific expression was identified by expressing deletion-reporter constructs in transgenic mice.[22] It is noteworthy that the mice generated in both of these studies were fertile. Thus, investigators should be able to use elements of these LHRH promoters to target expression of different genes to LHRH neurons *in situ,* and to propagate these transgenic mice.

III. The GT-1 Cells Have a Neuronal Phenotype

When the anterior hypothalamic tumor cells were dispersed and grown in culture, they were very heterogeneous.[12] Interestingly, the cells that expressed the mRNA for LHRH were observed to grow in clusters, were not very adherent to the tissue culture plates, and were often found floating in the medium. Over a period of six months, these less adherent cells were enriched through differential plating on plastic culture plates, such that a pure population of cells (GT-1) was obtained. These cells were later subcloned into the GT1-1, GT1-3 and GT1-7 cell lines.

All of the cells (both dividing and non-dividing cells) in each of the three GT-1 subclones stain for both LHRH and GAP.[12,18] By comparison, immunocytochemical and Northern blot analyses reveal that these cells do not contain tyrosine hydroxylase, somatostatin, corticotropin-releasing hormone, growth hormone-releasing hormone, or pro-opiomelanocortin.[12,19] Similarly, none of the GT1 subclones express genes that are normally found in astrocytes (glial fibrillary acidic protein) or in oligodendrocytes (myelin basic protein and myelin proteolipid protein). Instead, transcripts normally present in neurons (neuron-specific enolase and the 68 kDa neurofilament protein) and neuroendocrine cells (SNAP-25, VAMP-2, chromogranin B, and secretogranin I) are found in these GT1 cells.

The GT-1 cells not only express genes that are normally associated with neuronal function, but they also appear to possess a neuronal phenotype.[12,18,19] The cells have clearly defined perikarya and neurites (Figure 3a,b). Some of these neurites appear as short dendrite-like processes, while others can extend for several millimeters and are similar in appearance to the beaded axon-like processes of LHRH neurons *in vivo.*[6,23] These processes, on occasion, may also have growth cones and cilia at their tips (Figure 3e). Typically, the GT1-7 cells are the most "neuronal" in appearance, while the GT1-3 are the most rounded. The GT1-1 cells are intermediate in phenotype. It

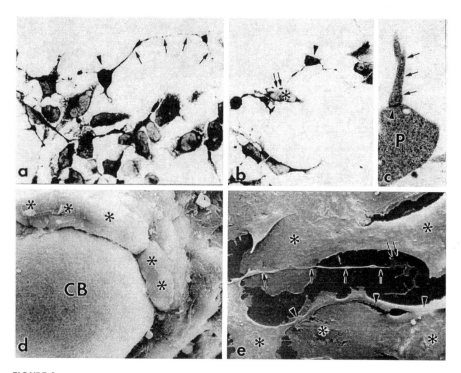

FIGURE 3

Morphology of GT1-7 neurons by light and electron microscopy. (a) A cluster of GT1-7 cells displaying LHRH immunoreactivity. One of the cells (arrowhead) possesses a long, varicose process (arrows). (b) A beaded neuronal process arising from a stellate immortalized cell (arrowhead) is juxtaposed (double arrows) next to another LHRH immunoreactive neuron (asterisk). (c) Low power electron micrograph of an immortalized LHRH neuronal process (arrows) terminating (arrowhead) on the perikaryon (P) of another GT1-7 cell. (d) Scanning electron micrograph of a cluster of GT1-7 neurons (asterisk) on the surface of a Cytodex bead (CB). (e) GT1-7 neurons (asterisks) captured by scanning electron microscopy. Outreaching thick (arrowheads) and thin (arrows) processes pass over neighboring cells. A growth cone is identified (double arrows).

should be noted that the appearance and response of these cells on the tissue culture plate can depend upon the source and lot of serum, the density of the cells on the plate, and on a variety of other factors.[24,25] In addition, a few investigators (personal communications) have noted that the neuronal phenotype of the GT1 cells can be lost when the cells are maintained in culture for prolonged periods of time.

Results from electron microscopy studies also indicate that the GT1 cells are neurons. The GT1-7 cells can be seen to have an abundance of ribosomes in the free and rough endoplasmic reticulum (Figure 4a), and the cytoplasm is densely packed with neurotubules (Figure 4b). In addition, secretory vesicles are often found adjacent to the Golgi complexes (Figures 4b,c,d) and are present in marginal positions in these cells.[19] Clathrin-coated pits and vesicles are frequently observed at the periphery of the cells. Omega profiles are also seen, indicative of active secretion from these cells. Both LHRH and GAP immunoreactivities are observed over the

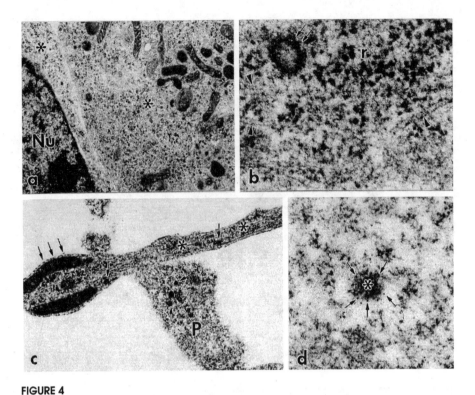

FIGURE 4

Electron micrographs of subcellular organelles in GT1-7 cells. (a) The cytoplasms of GT1-7 cells (asterisks) are rich in polyribosomes. Nu: nucleus of one of the cells. (b) Cytoplasmic detail of a GT1-7 neuron possessing ribosomes (r), neurotubules (arrowheads), and a single coated vesicle (double arrows). (c) A process (P) from one GT1-7 neurons intersects the process (asterisks) of another cell. The processes contain dense vesicles (arrows). The varicosity (multiple arrows) of the longer process contains two mitochondria. (d) A secretory vesicle (asterisk) tagged by colloidal gold particles (arrows) contains LHRH immunoreactivity.

ribosomes and they are found within (Figure 4d) and co-localized to the same secretory granules.[19] These latter data suggest that all of the processing steps from the conversion of pro-LHRH to LHRH probably occur within this organelle.

Although the GT1-7 cells appear as individual cells in culture, they are most often present as clusters (Figure 3a,b,d,e). Indeed, while the cells typically grow as a monolayer (Figure 3a,b), they can also grow on top of each other (Figure 3d), particularly when the cells reach ↓85% confluence. Even at low densities, the GT1-7 cells can be observed to make numerous contacts with each other (Figure 3a to e, 4c), and many of these same types of connections have also been described *in vivo*.[6,23] While many of these contacts are tight junctions, gap junctions and authentic synaptic connections (Figure 3c) can also be found.[26,27,28] These latter types of connections may be especially important in orchestrating and coordinating the secretion among the LHRH neurons.

These studies clearly show that the GT1 cells express a neuronal phenotype. Furthermore, these results provide a basis for establishing the GT-1 cell lines as an important neuronal model system that can be used in basic morphology studies.

IV. Use of Immortalized LHRH Cells to Study Neuronal Migration and Targeting

While the promoters used by Radovick and colleagues[11] and Mellon and co-workers[12] were responsible for cell-specific targeting, expression of the SV-40 T antigen led to the immortalization of these cells. In both studies, the offspring were infertile and the LHRH neurons in situ showed some unexpected behaviors. In the first case,[11] the immortalized neurons were confined to the olfactory bulb; the preoptic-hypothalamic region was devoid or depleted of LHRH neurons. In the second case, the preoptic-hypothalamic LHRH neurons failed to send their axons to the median eminence.[12,18] Recently, Radovick's[22] or Mellon's[21] groups have substituted a luciferase or choline acetyl transferase reporter gene for the SV-40 T-antigen and have found that the LHRH neurons in these mice migrated to the hypothalamus and that they apparently innervated the median eminance since their offspring were fertile. Taken together, these data indicate that the SV-40 T antigen may be interfering with the migration and/or targeting of the axons of the immortalized LHRH neurons. While the mechanisms that underlie these defects are unclear, several events may have contributed to their appearance. For instance, differentiation of LHRH progenitor cells in the olfactory placode could have been delayed such that by the time differentiation had occurred and migration begun, the cues or substratum for migration had disappeared. Alternatively, expression of the T-antigen might have impaired the ability of axons from the immortalized neurons to correctly target the median eminence. Finally, cell division might have inhibited both processes, or the T antigen might have suppressed certain genes associated with migration and axonal targeting.

The abilities of the axons of the GT1 cells to target the median eminence have been examined by several groups. In initial studies from our lab, we injected dispersed GT1-7 cells (~10 to 40×10^3 cells) either into the third ventricle or into the left or right medial basal hypothalamus (MBH) of hpg mice. In our hands, these conditions led to poor colonization of the hypothalamus. To enhance this process, GT1-7 cells were grown to approximately 90% confluency on 3 μm nucleopore polycarbonate inserts (Costar) in a 6-well tissue culture plate. Small clusters of cells were punched-out of the membrane with a blunt 20-gauge needle. This cell-membrane punch was implanted into the MBH by a stereotaxic instrument. Mice were sacrificed either three weeks (Figure 5) or two months later (Figure 6). The GT1-7 cells were found to readily colonize the MBH, and they assumed the typical LHRH morphology seen under in vivo conditions.[5-8] However, with very few exceptions, the processes of the GT1-7 neurons did not migrate into the median eminence.

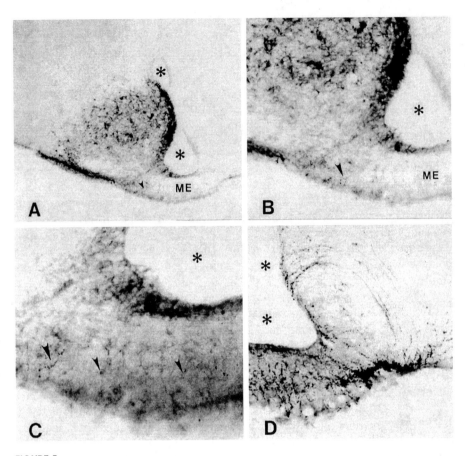

FIGURE 5

Colonization and migration of GT1-7 cells in hypothalamus of *hpg* mice. (A-C) GT1-7 cells were injected into the left side of the medial basal hypothalamus and *hpg* mice were sacrificed three weeks later. The neurons form a tumor that is well demarcated from the surrounding hypothalamic tissue. Except for one or two LHRH immunoreactive fibers (arrowhead), processes of the GT1-7 cells do not migrate into the median eminence. (D) LHRH immunostaining in the hypothalamus of an intact wild type mouse. Asterisks indicate the third ventricle. Magnifications: A (100x); B,D (200x); C (400x).

Similar deficits in targeting by the GT1-7 cells have also been observed Silverman, Gibson, and colleagues.[29,30] This deficiency appears to be unique to the GT1 cells since when fetal hypothalamic cells are implanted into the same brain region, almost all of the axons from the LHRH neurons project to the median eminence.[31]

Recently, another approach has been taken to study the role of the T-antigen in axonal targeting.[32] Hybrid genes were constructed containing the LHRH promoter and the coding region for either the T-antigen or a temperature-sensitive T-antigen. Expression of the temperature-sensitive gene should not be stable in the mouse (39°C), while it is stable at lower temperatures (34°C). As expected, all mice bearing the T antigen transgene were infertile. As noted previously,[12] although the LHRH neurons were able to migrate to the correct locations in the brain, no axon extensions were localized to the median eminence. Axonal targeting in other neuronal systems,

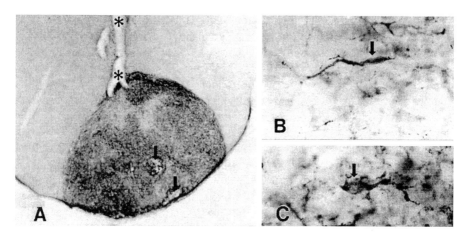

FIGURE 6

Morphology of a tumor of GT1-7 cells in the medial basal hypothalamus (MBH). (A) GT1-7 cells were implanted into the right MBH and *hpg* mice were sacrificed two months after surgery. The GT1-7 neurons form a tumor which occupies the entire right MBH, it partially invades the left MBH, and it is well demarcated from healthy hypothalamic tissue. The original organization of the MBH (arcuate nucleus, ventromedial nucleus, and median eminence) cannot be recognized. There are necrotic foci (panel A) in the tumor (arrows). Asterisks denote the third ventricle. (B-C) Examination of tumors from additional *hpg* mice. Within a tumor, numerous LHRH immunoreactive neurons are seen. These GT1-7 neurons exhibit the typical morphology ascribed to LHRH neurons under *in vivo* conditions (i.e., fusiform cells with only one prominent process). Magnifications: A (50×); B,C, (400×).

however, did not appear to be affected. By contrast, 6 of the 8 mice with the temperature-sensitive T-antigen were fertile. Taken together, these data suggest that the T antigen may interfere with the ability of the immortalized LHRH neurons to correctly target the median eminence.

In summary, inclusion of the 1.1 kb human promoter in the hybrid gene caused the migration of the LHRH neurons to be arrested. By contrast, 2.4 kb of the rat promoter in the hybrid gene permitted the LHRH neurons to migrate to the correct location within the brain, except now, their axons could not find their targets. These results suggest that there may be some elements on the human and rat promoters together with the T antigen that regulate targeting and migration of LHRH neurons. Identification of the transcription factors that bind these elements may provide some insights into these processes. Regardless, the GN or NLT and GT-1 cells should be adaptable to a variety of different formats where the molecular and cellular mechanisms that underlie neuronal migration and targeting can be investigated.

V. Studies of Interactions of GT-1 Neurons with Other Cells

The GT-1 cells are a homogeneous population of LHRH neurons. Due to this property, these immortalized neurons can be co-cultured with other cell lines or with primary cells from the pituitary, olfactory placode, hypothalamus, or other regions of the brain to study cell-cell interactions. A derivative of this technique has already been developed by Wray, Gainer and colleagues.[9,10] In this case, hypothalamic slice explant cultures were co-cultured with slices from the anterior pituitary. Immuno-cytochemical analysis revealed that the endogenous LHRH neurons preferentially migrated towards and invaded the pituitary. While direct innervation of the pituitary by LHRH neurons does not occur in mammals, it is normally observed in fish and some other species.[33]

In initial studies of pituitary-LHRH neuronal interactions, we co-cultured the GT1-7 neurons with either primary anterior pituitary cells or with αT3-1 immortal-ized gonadotropes.[34] Processes from GT1-7 cells extended towards the primary pituitary cells and the immortalized gonadotropes. Electron microscopy revealed that the GT1-7 cells established direct connections with both types of pituitary cells. The majority of these connections were through tight junctions. Furthermore, co-culture with the GT1-7 cells caused hypertrophy of the endoplasmic reticulum and an enhancement in the numbers of secretory vesicles in several of the anterior pituitary cell types and in the αT3-1 cells. These data indicate that the axons from the GT1-7 cells can target cells of the pituitary and that they can stimulate cellular responses which may be compatible with their role in regulating anterior pituitary function.

Besides examining pituitary-LHRH neuronal interactions, other investigators have investigated the role of glia in controlling LHRH neuronal function.[35-38] Initial studies[35] indicate that astroglial cells can stimulate the proliferation of the GT1-7 cells. Moreover, these astroglia stimulate LHRH secretion. Additional co-culture experiments[36-38] have indicated that astrocytes can secrete certain growth factors (transforming growth factor-β) or eicosanoids (prostaglandins) that can influence LHRH gene expression and secretion. Results from these studies emphasize the power of the co-culture approach in dissecting basic cellular interactive mechanisms that regulate LHRH gene expression and secretion. Thus, co-culture experiments with the GT-1 cells may provide some novel insights into the cellular and molecular mechanisms governing not only axonal targeting, but also intercellular communica-tion as it relates to reproduction and neurobiology.

Conclusions

The immortalized LHRH cell lines represent a unique tool for neuroscientists. Compared to most of the cell lines which are currently used in neurobiological studies, the lineages of the GN, NLT, and GT-1 neurons are well-known. Importantly, at least the GT-1 cells appear to mimic very closely the functional morphology of LHRH neurons *in vivo*. The availability of these cells for experimental purposes should greatly facilitate our understanding of the molecular and cellular mechanisms that form the foundations for neuronal migration, axonal targeting, and neuron-cell interactions that are so important not only to reproductive neuroendocrinology, but also to basic neuroscience.

Acknowledgments

We wish to thank Ms. JoAnn Reid for assistance with some of the electron micrographs and Dr. Andres Negro-Vilar (Laboratory of Molecular and Integrative Neuroscience, NIEHS, Research Triangle Park, NC 27709) for his support for some of this work. Finally, we wish to express our appreciation for the help that we have received from the many students and other co-investigators over the years.

References

1. McCann, S. M., Taleisnik, S., and Friedman, H. M., LH-releasing activity in hypothalamic extracts, *Proc. Soc. Exp. Biol. Med.*, 104, 432, 1960.
2. Matsuo, H., Baba, Y., Nair, R. M. G., Arimura, A., and Schally, A. V., Structure of the porcine LH- and FSH-releasing hormone I. The proposed amino acid sequence, *Biochem. Biophys. Res. Commun.*, 43, 1334, 1971.
3. Schwanzel-Fukuda, M. and Pfaff, D., Origin of luteinizing hormone-releasing hormone neurons, *Nature (London)*, 338, 161, 1989.
4. Wray, S., Gant, P., and Gainer, H., Evidence that cells expressing luteinizing hormone-releasing hormone mRNA in the mouse are derived from progenitor cells in the olfactory placode, *Proc. Natl. Acad. Sci. USA,* 86, 8132, 1989.
5. Silverman, A. J. and Krey, L. C., The luteinizing hormone-releasing hormone (LHRH) neuronal network of the guinea pig brain. I. Intra- and extrahypothalamic projections, *Brain Res.*, 157, 233, 1978.
6. Merchenthaler, I., Gorcs, T., Setalo, G., Petrusz, P., and Flerko, B., Gonadotropin-releasing hormone (GnRH) neurons and pathways in the rat brain, *Cell Tissue Res.*, 237, 15, 1984.
7. Silverman, A. J., Jhamandas, J., and Renaud, L. P., Localization of luteinizing hormone-releasing hormone (LHRH) neurons that project to the median eminence, *J. Neurosci.*, 7, 2312, 1987.
8. Merchanthaler, I., Setalo, G., Csontos, C., Petrusz, P., Flerko, B., and Negro-Vilar, A., Combined retrograde tracing and immunocytochemical identification of luteinizing hormone-releasing hormone- and somatostatin-containing neurons projecting to the median eminence of the rat, *Endocrinology,* 125, 2812, 1989.
9. Wray, S., Gahwiler, B. H., and Gainer, H., Slice cultures of LHRH neurons in the presence and absence of brainstem and pituitary, *Peptides*, 9, 1151, 1988.
10. Gainer, H., Kusano, K., and Wray, S., Hypothalamic slice-explant cultures as models for the long-term study of gene expression and cellular activity, *Regul. Peptides,* 45, 25–29, 1993.
11. Radovick, S., Wray, S., Lee, E., Nicols, D. K., Nakayama, Y., Weintraub, B. D., Westphal, H., Cutler Jr., G. B., and Wondisford, F. E., Migratory arrest of gonadotropin-releasing hormone neurons in transgenic mice, *Proc. Natl. Acad. Sci. USA*, 88, 3402, 1991.
12. Mellon, P. L., Windle, J. J., Goldsmith, P. C., Padula, C. A., Roberts, J. L., and Weiner, R. I., Immortalization of hypothalamic GnRH neurons by genetically targeted tumorigenesis, *Neuron*, 5, 1, 1990.

13. Mason, A. J., Pitts, S. L., Nikolics, K., Szonyi, E., Wilcox, J. N., Seeburg, P. H., and Stewart, T. A., The hypogonadal mouse: reproductive functions restored by gene therapy, *Science*, 234, 1372, 1986.

14. Mason, A. J., Hayflick, J. S., Zoeller, R. T., Young, W. S. III, Phillips, H. S., Nikolics, K., and Seeburg, P. H., A deletion truncating the gonadotropin-releasing hormone gene is responsible for hypogonadism in the *hpg* mouse, *Science*, 234, 1366, 1986.

15. Hanahan, D., Heritable formation of pancreatic β-cell tumors in transgenic mice expressing recombinant insulin/simian virus 40 oncogenes, *Nature*, 315, 115, 1985.

16. Windle, J., Weiner, R., and Mellon, P., Cell lines of the pituitary gonadotrope lineage derived by targeted oncogenesis in transgenic mice, *Mol. Endocrinol.*, 4,597, 1990.

17. Turgeon, J. L., Kimura, Y., Waring, D. W., and Mellon, P. L., Steroid and pulsatile gonadotropin-releasing hormone (GnRH) regulation of luteinizing hormone and GnRH receptor in a novel gonadotrope cell line, *Mol. Endocrinol.*, 10, 439, 1996.

18. Liposits, Z., Merchenthaler, I., Wetsel, W. C., Reid, J. J., Mellon, P. L., Weiner, R. I., and Negro-Vilar, A., Morphological characterization of immortalized hypothalamic neurons synthesizing luteinizing hormone-releasing hormone, *Endocrinology*, 129, 1575, 1991.

19. Weiner, R. I., Wetsel, W., Goldsmith, P., Martinez de la Escalera, G., Windle, J., Padula, C., Choi, A., Negro-Vilar, A., and Mellon, P., Gonadotropin-releasing hormone neuronal cell lines, *Front. Neuroendo.*, 13, 95, 1992.

20. Whyte, D. B., Lawson, M. A., Belsham, D. D., Eraly, S. A., Bond, C. T., Adelman, J. P., and Mellon, P. L., A neuron-specific enhancer targets expression of the gonadotropin-releasing hormone gene to hypothalamic neurosecretory neurons, *Mol. Endocrinol.*, 9, 467, 1995.

21. Lawson, M. L., Huang, K., and Mellon, P. L., Targeted reporter gene expression to GnRH neurons in transgenic mice by the minimal regulatory elements of the GnRH gene, in *Endocrine Society Meetings*, 1977, Vol. 79, 376 (abstract).

22. Wolfe, A. M., Wray, S., Westphal, H., and Radovick, S., Cell-specific expression of the human gonadotropin-releasing hormone gene in transgenic animals, *J. Biol. Chem.*, 271, 20018, 1996.

23. Liposits, Z., Setalo, G., and Flerko, B., Application of the silver-gold intensified 3,3'-diaminobenzidine chromagen to the light and electron microscopic detection of the luteinizing hormone-releasing hormone system in the rat brain, *Neuroscience*, 13, 513, 1984.

24. Wetsel, W. C., Eraly, S. A., Whyte, D. B., and Mellon, P. L., Regulation of gonadotropin-releasing hormone by protein kinase-A and -C in immortalized hypothalamic neurons, *Endocrinology*, 132, 2360, 1993.

25. Longo, K.M., Sun, Y., and Gore, A. C., "Maturation" of GT1-7 cells by neurotrophic factors, in *Endocrine Society Meetings*, 1997, Vol. 79, 377 (abstract).

26. Wetsel, W. C., Valença, M. M., Merchenthaler, I., Liposits, Z., López, F. J., Weiner, R. I., Mellon, P. L., and Negro-Vilar, A., Intrinsic pulsatile secretory activity of immortalized luteinizing hormone-releasing hormone-secreting neurons, *Proc. Natl. Acad. Sci. USA*, 89, 4149, 1992.

27. Matesic, D. F., Germak, J. A., Dupont, E., and Madhukar, B. V., Immortalized hypothalamic luteinizing hormone-releasing hormone neurons express a connexin 26-like protein and display functional gap junction coupling assayed by fluorescence recovery after photobleaching, *Neuroendocrinology*, 58, 485, 1993.

28. Charles, A. C., Kodali, S. K., and Tyndale, R. F., Intercellular calcium waves in neurons, *Mol. Cell. Neurosci.*, 7, 337, 1996.

29. Silverman, A. J., Roberts, J. L., Dong, K. W., Miller, M., and Gibson, M. J., Intrahypothalamic injection of a cell line secreting gonadotropin-releasing hormone results in cellular differentiation and reversal of hypogonadism in mutant mice, *Proc. Natl. Acad. Sci. USA*, 89, 10668, 1992.

30. Miller, G. M., Silverman, A. J., Roberts, J. L., Dong, K. W., and Gibson, M. J., Functional assessment of intrahypothalamic implants of immortalized gonadotropin-releasing hormone-secreting cells in female hypogonadal mice, *Cell Transplant,* 2, 251, 1993.

31. Silverman, A. J., Zimmerman, E. A., Gibson, M. J., Perlow, M. J., Charlton, H. M., Kokoris, G. J., and Krieger, D. T., Implantation of normal fetal preoptic area into hypogonadal mutant mice: temporal relationships of the growth of gonadotropin-releasing hormone neurons and the development of the pituitary/testicular axis, *Neuroscience*, 16, 69, 1985.

32. Morenter, S. M., Malcamp, C. A., Mason, A. J., Goldsmith, P. C., and Weiner, R. I., Temperature sensitive SV40 T antigen expression in GnRH neurons, unlike wild-type T antigen does not interfere with fertility in transgenic mice, in *Endocrine Society Meetings,* 1994, Vol. 76, 520 (abstract).

33. Schreibman, M. P., Halpern, L. R., Goos, H. J. Th., and Margolis-Kazan, H., Identification of luteinizing hormone-releasing hormone (LH-RH) in the brain and pituitary gland of a fish by immunocytochemistry, *J. Exp. Zool.,* 210, 153, 1979.

34. Liposits, Z., Wetsel, W. C., Reid, J. J., Merchenthaler, I., Mellon, P. L., and Negro-Vilar, A., Electron microscopic studies on co-cultures of immortalized hypothalamic LHRH neurons and primary or immortalized anterior pituitary cells. In *Society for Neuroscience Meetings,* 1994, Vol. 22, 17 (abstract).

35. Gallo, F., Avola, R., Costa, A., and Marchetti, B. (1993). Astroglial cells in primary culture release factors that promote the growth of GT1-7 LHRH neuronal cells and stimulate leukocyte proliferation. *Soc. Neurosci.,* 23:1634 (Abstract).

36. Melcangi, C., Galbiati, M., Messi, E., Piva, F., Martini, L., and Motta, M., Type 1 astrocytes influence luteinizing hormone-releasing hormone release from the hypothalamic cell lines GT1-1: Is transforming growth factor-β the principle involved?, *Endocrinology*, 136, 679, 1995.

37. Galbiati, M., Zanisi, M., Messi, E., Cavarretta, I., Martini, L., and Melcangi, R. C., Transforming growth factor-β and astrocytic conditioned medium influence luteinizing hormone-releasing hormone gene expression in the hypothalamic cell line GT1, *Endocrinology,* 137, 5605, 1996.

38. Ma, Y. J., von der Emde, K. B., Rage, F., Wetsel, W. C., and Ojeda, S. R., Hypothalamic astrocytes respond to transforming growth factor-α with the secretion of neuroactive substances that stimulate the release of luteinizing hormone-releasing hormone, *Endocrinology*, 138, 19, 1997.

Chapter 2

Assessment of Neuropeptide Receptor Gene Expression Using Reverse Transcriptase-Polymerase Chain Reaction (RT-PCR)

Janice H. Urban

Contents

I. Introduction

Molecular biology has greatly impacted the field of neuroendocrinology by facilitating the identification of receptor subtypes that otherwise might be difficult to detect using traditional receptor binding methodologies. Often for neuroendocrine systems, the relative abundance of gene expression for a particular receptor subtype can be very low. In particular, the application of reverse transcriptase-polymerase chain reaction (RT-PCR) to the field has allowed neuroendocrinologists the ability to amplify and quantitate low abundance signals. RT-PCR has the ability to detect 1 to 10 gene copies within a sample whereas other methods of quantitation such as *in situ* hybridization or nuclease protection assays detect 100 or 10^5 copies of RNA, respectively.[1] Another advantage of RT-PCR is that multiple transcripts, or alternatively spliced transcripts, can be measured within the same sample without pooling multiple samples.[2] Additional advantages to the use of RT-PCR are that it allows for the relatively quick processing of large numbers of samples and that optimal reaction conditions can be tightly controlled. However, the accuracy and reproducibility of this technique rely on the application and the development of proper controls.

This chapter will describe some of the considerations surrounding the design and optimization of a successful semi-quantitative or quantitative RT-PCR assay, and summarize some important concerns for reliable measurement of neuroendocrine parameters. While this technique can assess the expression of a number of different genes, the focus in this review will be on the expression of NPY Y1 receptor mRNA levels in the pituitary of the rat. We have identified the existence of the NPY Y1 receptor in the pituitary and using RT-PCR, have been able to reliably measure changes in receptor mRNA levels that occur under different steroid conditions in both male and female rats.

II. Reverse Transcriptase-Polymerase Chain Reaction Protocol

The development of the PCR reaction in the 1980s[3] has significantly expanded the development of many applications including gene cloning, identification of multiple or alternatively spliced RNA molecules, synthesis of gene probes, quantitation of gene expression, and the identification of human genetic diseases.[4-7] The power of PCR relies on its ability to amplify or produce large amounts of specific DNA fragments from a small amount of starting material or DNA template. Thus, the sensitivity of RT-PCR makes it an ideal technique to study the regulation of neuroendocrine genes, especially those expressed in low abundance. While this amplification is a definite advantage, the exponential nature of this process can amplify nonspecific products (contaminants), and slight variations in reaction conditions between samples can produce variable results within the same assay. Therefore, in order to have a reliable and reproducible assay it is imperative that attention be paid to the fundamentals of RT-PCR and the development of proper controls.

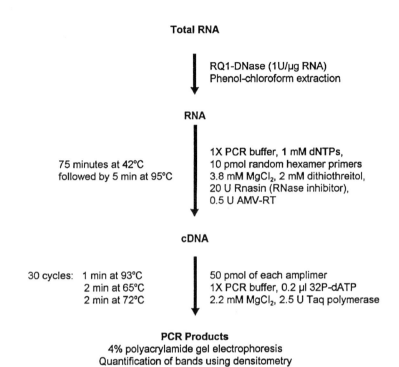

Total RNA

RQ1-DNase (1U/µg RNA)
Phenol-chloroform extraction

RNA

75 minutes at 42°C   1X PCR buffer, 1 mM dNTPs,
followed by 5 min at 95°C   10 pmol random hexamer primers
                                3.8 mM MgCl$_2$, 2 mM dithiothreitol,
                                20 U Rnasin (RNase inhibitor),
                                0.5 U AMV-RT

cDNA

30 cycles: 1 min at 93°C   50 pmol of each amplimer
            2 min at 65°C   1X PCR buffer, 0.2 µl 32P-dATP
            2 min at 72°C   2.2 mM MgCl$_2$, 2.5 U Taq polymerase

PCR Products
4% polyacrylamide gel electrophoresis
Quantification of bands using densitometry

FIGURE 1
Overview of RT-PCR protocol showing the different components to the procedure. RNA is isolated and treated with DNase to remove any source of contaminant DNA. The RNA is reverse transcribed in the presence of dNTPs, primers, buffers and AMV-RT. The resultant cDNAs are subjected to PCR with gene specific amplimers, buffers, ^{32}P-dATP, and Taq polymerase for 30 cycles of amplification. The PCR products are run on a 4% polyacrylamide gel, and band intensity is quantified using a phosphorimager.

An overview of the RT-PCR assay, represented by an outline of our protocol, is presented in Figure 1. Total RNA is extracted from tissues using the method previously described by Chomczynski.[8] RNA is treated with RQ1-deoxyribonuclease (DNase) and extracted using phenol-chloroform-isoamyl alcohol precipitation. The samples are reverse transcribed to cDNA using random primers and avian myeloblastosis RT (AMV-RT). Control reactions are included that contain the same reagents except that either AMV-RT or RNA is omitted from the samples. These controls check for the presence of DNA and non-specific contaminants, respectively. PCR is used to amplify a sequence of the NPY Y1 receptor cDNA and α-tubulin cDNA (which acts as an internal control for the efficiency of the reaction and the amount of RNA in each sample). Four dilutions of each sample are subjected to PCR to ensure that the amplification of the products occurs within a linear range. In our assay, these correspond to the equivalent of 0.125, 0.25, 0.5 and 1.0 µg of starting RNA. A ^{32}P-dATP label is incorporated during the PCR reaction so that the resultant products can be quantified using polyacrylamide gel electrophoresis and quantitative autoradiography. Each RT-PCR assay includes four dilutions of one RNA sample, and one sample from each experimental group is represented in an individual assay as an additional control on interassay variability. NPY Y1 receptor gene expression per sample is calculated as the average ratio of NPY Y1/α-tubulin for the consecutive dilutions. To maintain proper quality control, a control sample of pooled RNA is routinely run in each assay to determine inter- and intra-assay variation which is typically 9.5 and 11%, respectively. The following text will describe in more detail some of the technical considerations that are part of the optimization of this method.

A. RT Reaction

RT-PCR reactions start with RNA and require the sequential use of the reverse transcriptase and PCR reactions. During the RT process, RNA is reverse transcribed to cDNA which is then used as a template for PCR with oligonucleotide primers that are specific for the cDNA of interest. For RT reactions, especially as part of a quantitative assay, it is necessary to start with intact RNA and standardized RT reaction conditions as the amounts of cDNA produced by this reaction can vary between samples.

1. Preparation of RNA

A crucial step to a successful RT-PCR assay is the isolation of non-degraded RNA. At this point, and throughout the whole RT-PCR procedure, it is important to use RNase-free procedures and to minimize any possible sources of contamination (sterilize glassware, pipet tips, and solutions). RNA can be obtained from cells or tissue samples using standard RNA isolation procedures or a modified isolation method.[8] This rapid method for RNA isolation is adapted for the relatively quick processing of a large number of RNA samples. Using this method we can comfortably process 40 tissue samples in one day, and we obtain consistently uniform RNA quality and yields. The RT reaction can be performed on either total or poly(A)+ RNA, although it may be preferable to use total RNA as there is less variation in the quality and recovery from sample to sample. To verify the integrity of the RNA in each sample, an aliquot is run on a 1% agarose gel using RNase free conditions, stained with ethidium bromide and visualized under an ultraviolet lamp. Non-degraded total RNA will show two distinct bands representing the 18S and 28S ribosomal subunits; RNA of lesser quality will show a smearing pattern down the length of the gel.[9] Intense bands near the top of the gel indicate contamination by genomic DNA. This can be eliminated by the inclusion of an intermediate DNase step with RNase-free DNase (Promega, Madison, WI) to degrade any DNA which might interfere with the subsequent PCR amplification. Spectrophotometric readings at 260 and 280 nm will also indicate whether the RNA is intact and will more accurately quantitate the amount of RNA present. The total amount of RNA added to the RT reaction can vary, although 250 ng to 1 μg of total RNA should be sufficient for a single RT-PCR reaction.

2. cDNA Synthesis

The RT reaction consists of incubating RNA with small oligonucleotide sequences (6 to 30 base pairs), deoxynucleotides (dCTP, dGTP, dATP,dTTP) and a reverse transcriptase enzyme to produce complementary cDNA strands. Synthetic oligonucleotides anneal to the RNA forming a double-stranded complex which provides the substrate for the RT enzyme. These oligonucleotide sequences can either be (1) random hexamer primers that will anneal at random sites on the RNA, (2) oligo-dT (deoxythymidine) sequences which will specifically bind to the poly-(A)$^+$ tail of the mRNA or (3) sequences that are specific to the gene of interest, which will transcribe only those genes with that particular sequence. While all of these primers can perform equally well, it is generally held that using oligo-dT sequences at the 3' end of the molecule may be less efficient in producing a longer, full length cDNA strand, especially if a portion of the 5' end of the RNA is being amplified in the PCR. While the target specific primers are good for producing increased specificity for low abundance genes, the other primer selections (random hexamer primers, oligo-dT primers) have the advantage of allowing several RNAs to be reverse transcribed and ultimately, assayed within the same sample. The random hexamer primers are the most widely used especially when multiple genes are being analyzed or an exogenous control is used that does not contain a poly-(A)$^+$ tail.

Another component to consider in the RT step is the quality and amount of RT used. There are a number of RT enzymes available that have different activities; those most commonly used are the avian myeloblastosis virus (AMV-RT) and the Moloney leukemia virus (MMLV) reverse transcriptases. Both of these enzymes have minimal RNase H activity which can affect the efficiency of this step by degrading RNA. Enzyme preparations are available in which the RNase H activity is minimized thus resulting in longer cDNA strands. However, for most applications this intrinsic activity is not a problem. Yet, another variable that other investigators[2] have noted is that when small amounts of RNA are used, higher concentrations of AMV-RT can inhibit the subsequent PCR amplification even after heat denaturation of the enzyme. To prevent this from occurring, it may be necessary to titrate the amount of RT added per reaction. Variation can also be minimized in both the RT and PCR reactions if a master mix of the reagents is prepared and used for all samples in one assay.

B. PCR Reaction

The PCR reaction itself, consists of amplifying a portion of cDNA with gene specific primers and an internal (endogenous or exogenous) control with a heat stable DNA polymerase. The first cycle of PCR starts with a high temperature step (93 to 95°C) to denature DNA template and any previously synthesized sequences. The temperature is then lowered to facilitate annealing of the primers to the template; the temperature at this cycle is determined by the melting temperature (Tm) of the primers. Once this annealing step has occurred, the temperature is increased to 72°C (optimal for polymerase activity), and the DNA polymerase binds to the DNA duplex (primer-template) and synthesizes the DNA as defined by the two primer sequences. The DNA is denatured again and this process proceeds for a defined number of cycles. The kinetics of the PCR amplification also has a number of phases: the beginning cycles where the first few copies of DNA are being amplified; the exponential phase, which represents the linear amplification phase of the assay and finally the plateau phase where there is exhaustion of the enzyme and primers and essentially no more significant amplification of the product.[10] At this higher amplification phase there is more of a tendency for false bands to be generated. Therefore, it is critical to first test amplification of the desired products over a range of different cycle numbers. Other important aspects to consider for this reaction are the selection of primers, optimization of the PCR reaction (Mg^{++} concentration, amount of starting material, number of cycles for amplification) and the selection of an appropriate control.

1. Selection of Primers

The selection of good primers is essential to successful PCR. There are a number of points to good primer selection: (1) they should be between 18 to 23 base pairs and have very high homology and specificity to the target gene; (2) primers should not be complementary so they do not self-anneal. This is very important especially at the 3' ends of the molecule where primer-dimers can form; and (3) the dissociation temperature or melting temperatures (Tm) of the primers should be relatively the same. Computer software is available that helps in the design of primer sequences as well as provides information on the guanine/cytosine content, Tm of the primer pairs and the ability of the primers to anneal with each other. Each primer sequence should also be analyzed through GenBank (National Center for Biotechnological Information; NCBI; www.ncbi.nlm.nih.gov) to ensure that it has a unique sequence homology. When designing primers, the pairs should amplify a template sequence 0.200 to 1 kb in length and span an intron to allow amplification from RNA and contaminating genomic DNA to be distinguished by differences in size. For example, in the NPY Y1 receptor gene, there is an 80 bp intron between exons 2 and 3.[11] We have designed primers to be complementary to regions in exons 2 and 3; therefore a sequence amplified from the RNA will be 80 bp less than those that are amplified from genomic DNA. While this approach works well for a number of genes, it may be problematic when some coding sequences (especially G-protein coupled receptors) are not interrupted by an intron sequence, or there is the existence of pseudo-genes which also may not contain an intron. In this event, a DNase step prior to the RT reaction (see Figure 1) would help to degrade contaminating DNA. The amount of DNA contamination in PCR can be assessed using a control tube containing RNA with no reverse transcriptase. Since PCR will not amplify RNA, any bands resulting in this control sample will be the result of a contaminant (DNA).

In many RT-PCR reactions, multiple primer pairs (more than one pair) may be added to amplify different genes such as the target gene and the control gene. Some of the issues that arise with the use of multiple primers is that one pair may interfere with the efficiency of the others, the primers themselves may anneal with each other, and if the Tms are different, there could be an increase in non-specific product generation if the annealing temperature is too low for one set. If the reaction is designed for multiple primer pairs, it is important to compare the sequences and check for overlap or complementarity, annealing properties (they should have a similar Tm) and ensure that they do not interfere with other RT-PCR product formation. When running multiple primer pairs in a single reaction, each of the primers should first be run separately, and then combined to compare how the kinetics of the reaction may change when the primers are combined. Finally, the concentration of the amplimers (PCR primers) should be titrated so that there is a high enough concentration to allow the PCR amplification to proceed without causing unnecessary primer-dimer formation. Both of the primers should be added in a similar concentration, and depending upon the template and complexity of the DNA, the range of final primer concentration in PCR should be between 0.1 to 0.5 μM (8 to 50 pmol per 100 μl volume).

2. Optimization of Assay Conditions

There are many components to the PCR aspect of this protocol. While the primers are important in obtaining the necessary specificity of the assay, the pH and buffer conditions such as Mg^{++} concentration are also of paramount importance. Mg^{++} concentrations can directly affect the yield and amplification of the DNA product. Higher concentrations of Mg^{++} can stabilize the DNA and prevent the complete dissociation of the product during the denaturation step thus inhibiting further amplification.[10] Conversely, lower Mg^{++} concentrations can inhibit the correct amplification of the template and result in lower yields of the DNA product. When using new primer sets, it is helpful to run concentration curves using different Mg^{++} concentrations in the RT-PCR assay. While for most of our primer sets, the concentration of Mg^{++} remains fairly similar, we have noted that a shift in the concentration of as little as 0.2 mM can cause a significant decrease in DNA amplification.

Additionally, the pH may affect PCR amplification with some primers and templates. Generally, the buffers supplied with the polymerase enzymes are Tris-based and maintain a pH within the range of 7.0 to 8.0. While we have not found this parameter to be of concern in our system, it may be important to check this in other systems. Optimization of these variables can be tedious. However, currently there are a number of kits commercially available that allow easy, standardized, optimization of both Mg^{++} concentrations and pH levels.

3. Annealing Temperature and Determination of Cycle Number

The annealing temperature (melting temperature; Tm) of a DNA sequence refers to a point where approximately one-half of the amplimers are annealed to the target DNA. This number (Tm) can be empirically defined or it can be obtained from the computer programs used to identify the primer sequence or provided by the biotechnology company that supplies the oligomers. When setting up PCR it is important to have annealing temperatures that are optimized for the specific primers or extraneous bands result if the temperature is too low, and inadequate annealing occurs if the temperature is too high. Generally, temperatures between 55 to 65°C work well for primers 18 to 22 bp in length with a guanine/cytosine content of 50%.

Once these conditions are set for RT-PCR, other variables to consider are the amounts of RT product needed to detect a signal and the number of PCR cycles to be preformed to produce linear and reproducible samples. These aspects are crucial to consider especially when the assay is needed to quantitate mRNA levels. In neuroendocrine systems, the expression of a gene can sometimes vary greatly depending upon the physiological condition. For example, with respect to gonadotropin hormone releasing hormone (GnRH) receptor mRNA levels in the pituitary, basal expression is relatively low until the afternoon of proestrus when the levels of GnRH receptor mRNA are acutely increased in preparation of the preovulatory surge and as a result of steroid priming.[12] In some of our previous experiments we have found NPY Y1 receptor mRNA levels to change by as much as 10-fold depending upon the stage of the estrous cycle.[13] These physiological examples therefore make it imperative that the RT-PCR assay be dynamic to accurately quantify low and high levels of gene expression. Under these conditions, the amount of RNA added per sample as well as the number of PCR cycles, need to be able to amplify the products within a linear range. To assess the number of cycles necessary, samples are titrated through a range of different cycles. Typically, for neuroendocrine systems, 25 to 30 cycles is optimal; if the reaction is allowed to continue for more than 30 cycles nonspecific artifacts may be generated, thus interfering with the assay. Also with a higher number of cycles, it is easy to reach the plateau phase of amplification where the template is no longer amplified linearly. Another consideration that helps ensure the samples are amplified within a linear range is the use of multiple dilutions of the same sample. We generally reverse transcribe 5 μg of total RNA and then use four serial dilutions of the RT product to cycle through the PCR. Using four dilutions allows us to verify that each sample is being read within the linear range (Figure 2), and that if dramatic changes in gene expression occur, they can be read on either the high or low end of the "mini-curve".

C. Controls

Using RT-PCR in quantitative measurements necessitates the use of an internal control. The controls are co-amplified in the same samples as the target gene and monitor variations in the RT-PCR reactions as well as serve as standards for quantitation. There are two types of controls that are used for the quantitation of RT-PCR products: endogenous and exogenous standards. Endogenous controls are those molecules that are already present within the sample, whereas exogenous controls are synthetic cRNA or cDNA molecules that are added in known quantities to the RT-PCR mix. The following discussion will outline some of the advantages and disadvantages to using these different controls.

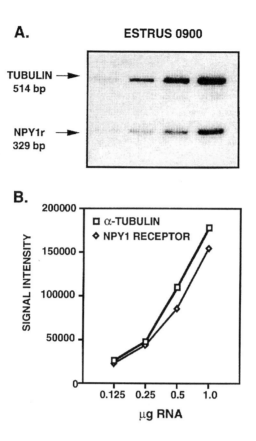

A. ESTRUS 0900

TUBULIN →
514 bp

NPY1r →
329 bp

B.

FIGURE 2
Results of a quantitative RT-PCR for NPY Y1 receptor mRNA in the anterior pituitary of a female rat on estrus day. (A) Autoradiogram of PCR products generated from 0.125, 0.25, 0.5, and 1.0 μg RNA starting material. NPY Y1 receptor is represented as a band of 329 bp, a faint band indicating the presence of genomic DNA is observed below the tubulin band at 509 bp. (B) Quantitation of the signal intensities from the autoradiogram in A. Both the NPY Y1 receptor and tubulin signals show increased intensities with an increase in RNA content, and these signals remain linear and parallel over the range of concentrations.

1. Endogenous Controls

Endogenous internal standards consist of amplifying the target gene with an endogenous RNA present within the sample. Typically, these endogenous controls are "housekeeping genes" and may be molecules such as metabolic enzymes (G3PDH), structural genes, (actin or tubulin), or ribosomal proteins. The reason these compounds are chosen is because they are ubiquitous and can be assessed in different tissue samples, and their constitutive expression overall will not change dramatically with experimental treatments. The advantages to using endogenous controls are that they will control for sample to sample variation in the kinetics of the RT-PCR mix and control for the amount of RNA added to the sample and the amount of degradation that occurs within the first stage of the assay. Using these controls to measure RNA levels requires the co-amplification of the target gene with the endogenous standard. The target gene is normalized to the internal control and the samples can then be compared to each other. A number of studies have used this method to assess changes in different genes associated with neuroendocrine control.[2,12-14]

As with most aspects of RT-PCR, there are several cautions that should be heeded when designing controls for RT-PCR. One important concern is that with most of the housekeeping genes, their levels of gene expression may vary and therefore the selection of the internal standard may depend upon the application for which it is used. Other concerns have to deal with the relative abundance of the standard to the target gene. Both of the targets should have a similar abundance so that when the products are amplified, they will have similar kinetics and show overlap at the exponential phase of the assay. If one product amplifies quicker, the primer pair for this gene can be added to the tubes at a later time (two-step PCR). Also, the length and optimal annealing temperature for both primer sets should be similar so that the likelihood of non-specific artifacts can be minimized. It is necessary when using multiple primer pairs to optimize the PCR with respect to the amount of starting material and the number of cycles used so that the relative ratios of the target to control gene can be compared. Using four different dilutions of the RT product ensures that we obtain a linear signal for both the α-tubulin and NPY Y1 receptor RNA at a constant of 30 cycles of PCR (Figure 2). The use of different RT product dilutions is important in our experiments as the expression of NPY Y1 receptor and α-tubulin are almost in a 1:1 ratio in one experimental condition and 0.1:1 in another. These dramatic changes in NPY Y1 receptor mRNA are due specifically to changes in Y1 receptor gene expression as we do not see significant changes occurring in α-tubulin expression under different conditions. While endogenous controls are useful for measuring changes in RNA levels under different physiological conditions, it is not well suited for determining the amounts of RNA in a quantitative manner; this method of using endogenous controls is often referred to as semi-quantitative RT-PCR. The following section on the use of exogenous standards will take issue with the use of more absolute, quantitative RT-PCR methods.

2. Exogenous Controls

The use of exogenous controls in RT-PCR to directly quantify the resultant products is often referred to as competitive or quantitative PCR. This method relies on the spiking of individual reactions with different concentrations of a control cRNA or cDNA template. The template is then co-amplified with the target gene, usually with the same primers, and the concentration of the target can be interpolated from the dilutions of the exogenous control.[15-17] When these dilutions are co-amplified with the target, the point at which both resultant products are equal is when the target concentrations are identical to the exogenous control. Typically, the control used in these reactions has the exact same primer sequence as the target gene. Usually, the control template has the same sequence as the target, only there is a deleted, or added, sequence at a given restriction site. These mutations allow the investigator to readily distinguish the template from the control based on a size difference. The advantage of this is that the primers compete for the target and control RNA similarly. Using the same sequence for the control assumes that it will have the same amplification kinetics as the target and therefore, the issues surrounding the use of multiple primers (as mentioned above) is obviated. As with the endogenous control, it may be important to quantify the signal within the linear phase of the PCR[17] while other investigators have been able to quantify the signal within the non-linear or plateau phase of PCR.[18] One consideration with measuring expression in the non-linear phase is that since there is strong homology between the amplified sequences (control and template), heteroduplexes may form thereby obscuring the actual relationship of the template to control. While this methodology is more tedious than using an endogenous control, it does allow you to obtain more of an absolute value for RNA concentrations within the sample.

Finally, whether using an endogenous or exogenous standard there are options to the method of detection. Incorporation of a radiolabeled nucleotide such as ^{32}P-dATP into the PCR mix allows a greater sensitivity and ease of quantitation either by gel electrophoresis and quantitative autoradiography, or phosphorimaging, or by excising the band and using a scintillation counter to assess the amount of radioactivity. The products can also be analyzed by Southern blotting with a radiolabeled probe. Currently, different non-isotopic methods are also used which take advantage of ethidium bromide staining in the gel,[16] fluorescent and chemiluminescent[19] methodologies to assess the amounts of DNA generated via RT-PCR.

III. Uses of RT-PCR

We have developed an RT-PCR assay to specifically assess changes that occur in NPY Y1 receptor mRNA levels within the pituitary during the estrous cycle of the female rat and under controlled gonadal-steroid conditions in the male rat. By considering many of the issues presented in this review article, the assay that developed (refer to Figure 1) is very reproducible as assessed by the inter- and intra-assay variation controls that are routinely run in the individual assays. Figure 3 shows representative data demonstrating the relative expression of NPY Y1 receptor mRNA levels in the anterior pituitary of male rats that were sham-operated (intact), castrated or castrated and replaced with physiological levels of testosterone (T). Two weeks following castration, there is a significant decrease in NPY Y1 receptor gene expression. Replacement with T reversed this effect suggesting that NPY Y1 receptor gene expression is regulated in part by testosterone. We can measure NPY Y1 receptor gene expression in the pituitary using Northern analysis, but with low levels of receptor gene expression, the process is long and not easily adapted to processing a large number of samples. This is one example of how the RT-PCR technique can be adapted to those systems that have low abundance mRNA levels or have cells dispersed over a wide area. RT-PCR has been used to measure changes in gene expression of a number of different neuropeptide receptors including GnRH receptor,[12] growth

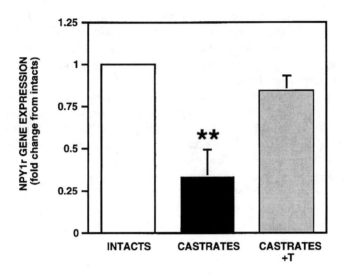

FIGURE 3

NPY Y1 receptor mRNA levels in the anterior pituitary from male rats that were intact (sham-operated), castrated (2 weeks), or castrated and received physiological implants of testosterone. Data are represented as mean ± S.E.M. n = 3 to 4. NPY Y1 receptor mRNA levels are expressed as a fold change from the intact control group. **significantly different from all other groups, $p < 0.01$; one way ANOVA followed by a Tukey's post hoc test.

hormone releasing hormone receptor,[14] angiotensin II receptor,[20] and interleukin-6 receptor.[19] Other reports have used competitive RT-PCR to measure mRNA levels in micropunches of the hypothalamus,[21] a technique which is widely used in neuroendocrinology. While this technique may lack anatomical resolution of some other methods, such as *in situ* hybridization, the advent of *in situ* RT-PCR may resolve some of these issues.[22] No one technique can yield a definitive answer regarding the regulation of neuroendocrine systems. The method of RT-PCR, however, can reliably and reproducibly measure the steady state levels of gene expression in a large number of samples. One of the advantages of this technique is that within a small sample size, the levels of many different genes can be assessed while with other methods, this simultaneous measurement can prove to be quite difficult. As molecular biological methods have greatly benefitted a number of fields, RT-PCR has recently gained acceptance within the fields of endocrinology and neuroendocrinology as a method that can reliably assess changes in a number of neuropeptide/neurotransmitter systems.

Acknowledgments

This work was supported in part by American Heart Association-Metropolitan Chicago Grant-In-Aid and NIH R29 MH53663. The author would like to thank Dr. Angela Bauer-Dantoin for the critical reading of this manuscript.

References

1. Piatek, M., Jr. Luk, K.C., Williams, B., and Lifson, J.D., Quantitative competitive polymerase chain reaction for accurate quantitation of HIV DNA and RNA species, *Biotechniques*, 14, 70, 1993.
2. Foley, K.P., Leonard, M.W., and Engel, J.D., Quantitation of RNA using the polymerase chain reaction, *Trends Genet.*, 9, 380, 1993.
3. Mullis, K.B., The unusual origin of the polymerase chain reaction, *Sci. Am.*, April, 56, 1990.
4. Lenstra, J.A., The application of the polymerase chain reaction in the life sciences, *Cell. Molec. Biol.*, 41, 603, 1995.
5. Eidne, K.A., The polymerase chain reaction and its uses in endocrinology, *Trends Endocrinol. Metab.*, 2, 169, 1991.
6. Ronai, Z. and Yakubovskaya, M., PCR in clinical diagnosis, *J. Clin. Lab. Anal.*, 9, 269, 1995.
7. Lovenberg, T.W., Liaw, C.W., Grigoraidis, D.E., Clevenger, W., Chalmers, D.T., De Souza, E.B., and Oltersdorf, T., Cloning and characterization of a functionally distinct corticotropin-releasing factor subtype from rat brain, *Proc. Natl. Acad. Sci. USA*, 92, 836, 1995.
8. Chomczynski, P., A reagent for the single-step simultaneous isolation of RNA, DNA and proteins from cell and tissue samples, *Biotechniques*, 15, 532, 1993.
9. Dumas Milne Edwards, J.B., Ravassard, P., Icard-Liepkalns, C., and Mallet, J., cDNA cloning by RT-PCR, in *PCR 2: A Practical Approach*, McPherson, M.J., Hames, B.D., and Taylor, G.R., Eds., Oxford University Press, New York, 1995, chap 6.
10. Kidd, K.K. and Ruano, G., Optimizing PCR, in *PCR 2: A Practical Approach*, McPherson, M.J., Hames, B.D., and Taylor, G.R., Eds., Oxford University Press, New York, 1995, chap 1.
11. Eva, C., Oberto, A., Sprengel, R., and Genazzani, E., The murine NPY-1 receptor gene: structure and delineation of tissue-specific expression, *Fed. Eur. Biochem. Soc.*, 314, 285, 1992.
12. Bauer-Dantoin, A.C., Hollenberg, A.N., and Jameson, J.L., Dynamic regulation of gonadotropin-releasing hormone receptor mRNA levels in the anterior pituitary gland during the rat estrous cycle, *Endocrinology*, 133, 1911, 1993.
13. Urban, J.H. and Levine, J.E., Neuropeptide Y1 receptor gene expression during the rat estrous cycle, *Soc. Neurosci.*, 21, 112.4, 1995.
14. Miller, T.L. and Mayo, K.E., Glucocorticoid regulation of pituitary growth hormone releasing hormone receptor mRNA expression, *Endocrinology*, 138, 2458, 1997.
15. Zamorano, P.L., Mahesh, V.B., and Brann, D.W., Quantitative RT-PCR for neuroendocrine studies, *Neuroendocrinology*, 63, 397, 1996.
16. Santagati, S., Bettini E., Asdente, M., Muramatsu, M., and Maggi, A., Theoretical considerations for the application of competitive polymerase chain reaction to the quantitation of a low abundance mRNA: estrogen receptor, *Biochem. Pharmacol.*, 46, 1797, 1993.
17. Wang, A.M., Doyle, M.V., and Mark, D.F., Quantitation of mRNA by the polymerase chain reaction, *Proc. Natl. Acad. Sci. USA*, 86, 9717, 1989.

18. Gilliland, G., Perrin, S., Blanchard, K., and Bunn, H.F., Analysis of cytokine mRNA and DNA: detection and quantitation by competitive polymerase chain reaction, *Proc. Natl. Acad. Sci. USA*, 87, 2725, 1990.

19. Gadient, R.A. and Otten, U., Differential expression of interleukin-6 (IL-6) and interleukin-6 receptor (IL-6R) mRNAs in rat hypothalamus, *Neurosci. Lett.*, 153, 13, 1993.

20. Llorens-Cortes, C., Greenberg, B., Huang, H., and Corvol, P., Tissular expression and regulation of type 1 angiotensin II receptor subtypes by quantitative reverse transcriptase-polymerase chain reaction analysis, *Hypertension*, 24, 538, 1994.

21. Kim, K., Jarry, H., Knoke, I., Seong, J.Y., Leonhardt, S., and Wuttke, W., Competitive PCR for quantitation of gonadotropin-releasing hormone mRNA level in a single micropunch of the rat preoptic area, *Mol. Cell. Endocrinol.*, 97, 153, 1993.

22. Bagasra, O. and Pomerantz, R.J., Detection of HIV-1 in brain tissue of individuals with AIDS by *in situ* gene amplification, in *PCR in Neuroscience*, Vol. 26, Sarkar, G., Ed., Academic Press, San Diego, 1995, chap. 23.

Chapter **3**

Identification of Biotinylated Ligands on Specific Target Cells in the Pituitary: Studies of Regulation of Binding

Gwendolyn V. Childs

Contents

0-8493-3363-6/98/$0.00+$.50
© 1998 by CRC Press LLC

I. Introduction

Affinity cytochemistry is that branch of cytochemistry that detects a ligand bound
to a target site or cell. Sometimes the term is used to refer specifically to the detection
of ligands bound to receptors. A number of approaches can be used in the labeling
or detection system. One can label the ligand with a radioactive detector and then
apply it to the cells. The detection system for the radioactive compound is autoradiography. Its advantage is its sensitivity and the fact that one can label the ligand
directly. However, a major disadvantage of this system is spatial resolution, that is,
the site of the reaction product (silver grains) does not always correspond to the
exact site of the ligand.

Alternatively, one can use a heavy metal label, such as colloidal gold or ferritin
to label the ligand. The advantage is the ease of detection with this direct labeling
method. However, the relatively large size of the label may interfere with the binding
affinity of the ligand or completely prevent it from binding.

Biotin is a small molecule, about the size of an amino acid. Thus, it is less
likely to interfere with the affinity of a ligand. It was used in pioneering work by
Bayer et al.[1] to label ligands. The biotinylated ligand can then be detected by systems
involving a link with avidin which has a high affinity for biotin (10^{-15} M). Avidin
can in turn be linked to enzymes, fluorescein, colloidal gold or other signaling
molecules. This provides a high affinity detection system that can be used to detect
the site of attachment of a ligand.

We have experimented with biotinylated ligands since 1983.[2-22] They are used
to detect receptors in anterior pituitary cells as well as other cell types. The protocol
begins by exposing cells to the biotinylated ligand. Then they are fixed, or cooled
to 4°C, depending on the detection system to be used. The biotinylated ligand is
then detected by avidin peroxidase, avidin fluorescein, or avidin linked to gold or
ferritin.[2-22] In the earliest studies we used the Avidin-biotin peroxidase kit developed
by Hsu et al.[23]

In these studies, we learned that one advantage of this system is that it can
detect biotinylated ligands used in physiological concentrations.[2-9] The system can
also be used to detect cells while they are living.[10,11] Furthermore, it can be used at
both the light and electron microscopic levels. It is quantifiable either by densitometry or particle counting (in the case of colloidal gold labels). Finally, our earliest
studies showed that it can be combined with immunolabeling systems that allow
further identification of the target cells.[3] The following presentation will focus on
the methods used in this laboratory for the detection and quantification of biotinylated ligands bound to pituitary target cells. The presentation will also show how
this can be combined with immunocytochemistry to allow identification of target
cells. In addition, examples will be presented to show how this system can be used
to study regulation of receptor expression.

II. Materials and Methods

A. Collection and Plating of Pituitary Cells

The labeling and detection systems depend on healthy populations of anterior pituitary cells that are dispersed and either grown in monolayer cultures, or left in suspension (for electron microscopic studies). The labeling thus depends upon an interaction between the biotinylated ligand and the living cell.[15-22] The time between dispersion and actual labeling varies with the experimental conditions. For example, we may want to test the effects of 24 to 48 hr incubation in steroids.[5,12] Or, tests of the receptor population could be extended to include the effect of culture conditions themselves.[17] To test the effects of the growth media on the receptor populations, we always compare labeling in a freshly dispersed culture with that grown for more extended periods.[12]

Male or female Sprague-Dawley rats are housed three per cage under artificial illumination between 0600 hr and 1800 hr and given food and water *ad libitum*. They are acclimated for 1 week before they are used in any experimental protocol. The animal care and use protocol is approved annually by the Institutional Review Committee.

The rats are sacrificed by guillotine and the anterior pituitary glands are collected after removal of the neurointermediate lobe. They are rapidly placed in cold Dulbecco's modified Eagle's medium (DMEM) (JRH Biosciences Lenexa, KS) containing 0.3% bovine serum albumin (BSA, Sigma Chemical Co. St. Louis, MO), 1.8 g/500 ml HEPES (Sigma Chemical), and 24.65 ml/500 ml sodium bicarbonate (JRH Biosciences). In order to prevent bacterial growth, 1 μl/100 ml gentamicin (Sigma Chemicals) is used. The dissociation protocol is performed as reported previously with 0.3 to 0.4% Trypsin for 15 min.[14] The pituitary pieces are placed in the trypsin on a slow shaker for 15 min and then resuspended in a solution containing trypsin inhibitor and DNase.[14] The pituitary cells are then dissociated by passages through a Pasteur pipette or syringe with an 18 gauge needle 40 to 50 times. After each set of 10 to 20 passages, the supernatant is transferred to a separate vial and more trypsin inhibitor is added to the remaining pieces. After all pieces have been dissociated, the cells are spun at 900 rpm and resuspended in fresh DMEM. They are either plated, or used in suspension for the electron microscopic detection of receptors.

Tests have been made of the effect of trypsin on the receptor population in general. All ligands tested will bind to freshly dispersed pituitary cells.[2-4] Furthermore, there is no increase in binding if the cells are grown for 12 hr. This suggests that the concentration of trypsin used does not significantly reduce receptivity. In addition, cells are tested for viability by the trypan-blue dye exclusion test. The protocol produces 2 to 4 million cells per pituitary that are 98% viable.

After dispersion, cells are suspended in DMEM containing 0.005 mg/ml insulin (Sigma Chemicals), 0.05 mg/ml transferrin (Sigma Chemicals), and 0.001 mM sodium selenite (Johnson Matthey Chemical Ltd., New York, NY.) They are then plated onto glass coverslips (AH Thomas Sci., Swedesboro, NJ) that are coated with poly-D-lysine (Sigma Chemicals). These 13 mm coverslips are housed in 24 well trays. The plating density depends on the cell population being studied and ranges from 5,000 cells per well (for enriched corticotrope populations) to 40,000 to 50,000 cells per well for mixed pituitary cell populations. The cells are plated at 37°C for 1 hr and then fed with the same DMEM. Some cultures are also grown in 10% fetal bovine serum (JRH Sciences); however, some experiments have used serum-free media, especially if tests of regulatory peptides are being conducted. The cells are grown in DMEM alone, or in media containing test peptides or steroids. After the designated plating period, the cultures are stimulated with the biotinylated ligand, as described below. Each ligand is tested for its biological activity, sensitivity, efficiency and specificity as a labeling tool. These tests will be described in the following paragraphs. The production of the biotinylated ligands being used in these studies is described in the References 2-4, 7, 9, 15, 22, 23, 24, and 25.

B. Biological Tests of the Biotinylated Ligands

During the past 14 years, our laboratory has collaborated with biochemists to produce biotinylated releasing hormones that also retain biological activity.[2-4,7,9,15,22] This activity is defined both by binding affinity and also the capacity to release the target cell hormones. In the first set of experiments with gonadotropin releasing hormone (GnRH), the attachment of biotin to a lysine substitution in the 6th position actually made the ligand more potent in its ability to release luteinizing hormone (LH) or follicle stimulating hormone (FSH) from cultured pituitary cells.[2,3,26] Its potency was compared with that following exposure to unlabeled D-Lys[6]-GnRH. The comparison of tests of this ligand is presented in References 2 and 3 and summarized in Reference 26. More recent studies of biotinylated ligands have shown however that certain conjugation methods will attach multiple biotins to GnRH[22,24,25] which reduces potency. Therefore, Dr. B. Miller and his colleagues have used HPLC to characterize and separate potent monobiotinylated products.[22,24,25]

Another approach conjugated biotin to an amino acid and then added the derivative during peptide synthesis. This last approach was tried because a simple conjugation of biotin to corticotropin releasing hormone (CRH) added at least three biotins to the releasing hormone, resulting in a product that was not as biologically active as its native counterpart.[4] One of the lysines bound by biotin was needed for biological activity. Therefore, biotin was first attached to serine which was then added during the synthesis of CRH.[9] The resulting ligand had biotin attached only to the N-terminal serine.[9] This biotinylated CRH was fully active as a secretagogue for adrenocorticotropin when tested in our pituitary cell bioassays.[9,26]

The tests for biological activity involve pituitary cells plated for 3 to 4 days as described in the first section.[2-5,9] They are exposed to 0, 0.1, 0.5, 1, 5, 10 and 50 nM biotinylated ligand or non-biotinylated ligand for 3 to 4 hr at 37°C. The diluent for the ligands is DMEM including 10^{-4} M ascorbic acid (as an antioxidant) and 100 K IU aprotinin as a protease inhibitor.[2-5,9] At the end of the exposure time, the media are collected and assayed by radioimmunoassay for the target cell hormones. One of the latest ligands now being tested is a monobiotinylated growth hormone

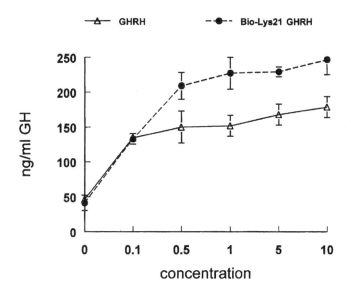

FIGURE 1

Bioassay for growth hormone secretion from 3-day cultures of female rat pituitary cells stimulated for 3 hr with either growth hormone releasing hormone (GHRH) or monobiotinylated GHRH (Bio-Lys[21] GHRH). The attachment of biotin to the GHRH has not interfered with its ability to stimulate the release of growth hormone. The growth hormone was assayed by radioimmunoassay with the National Pituitary Agency RIA kit.

releasing hormone (GHRH). An example of a comparative test of biotinylated and non-biotinylated GHRH is shown in Figure 1. In most tests, it is as potent a secretagogue for GH as is the non-biotinylated GHRH. The test in Figure 1 shows a higher level of GH released following exposure to the biotinylated GHRH.

C. Cytochemical Tests of the Biotinylated Ligands

1. Establishing the Cytochemical Detection System

At the time of the experiment, cell monolayers (for light microscopic studies) or cell suspensions (for electron microscopic studies) are washed gently in DMEM and exposed to optimal concentrations of the biotinylated ligand for optimal times (see tests below). The diluent is the same buffer described for the bioassays in the above paragraphs. The cells are then fixed in 2.5% glutaraldehyde (for light or electron microscopic immunoperoxidase studies) or 4% paraformaldehyde (for fluorescence studies). The glutaraldehyde is diluted in 0.1 M phosphate buffer containing 4.5% sucrose. p-Formaldehyde is diluted in 0.1 M phosphate buffer. After a 4× phosphate buffer wash (including 4.5% sucrose), the cells are treated with serum proteins to block non-specific binding sites. This can include 1 to 5% bovine serum albumin, 5 to 10% normal goat serum (used in immunolabeling) or 1 to 2% non-fat dry milk (or a combination of these). This blocking step lasts for 15 min at room temperature.

For routine light microscopic studies, the avidin detection system is applied for 60 min.[2-5] Most frequently this detection system has been the avidin biotin peroxidase complex (ABC elite) kit (Vector Laboratories, Burlingame, CA).[2-9,12,14,17-22] However, for the electron microscopic studies, we have also used streptavidin linked to ferritin, or colloidal gold.[6,8,9,16,20] Finally, we have used streptavidin fluorescein with paraformaldehyde fixed cells.[9] If cell sorter grade streptavidin fluorescein is used and the cells are cooled to 4°C, then the ligand can be detected before fixation on living cells.[9-11,13] This allows electrophysiologists to identify target cells and also record activities of ion channels on identified cells.[10,11]

The approach with living cells will also allow studies of secretion from individual target cells with the use of the Reverse hemolytic plaque assay (RHPA[10]). After the avidin fluorescein is applied for 15 min at 4°C, the cells are immediately exposed to the protein A conjugated red blood cells, the complement, and the specific antisera to the hormone secreted by the target cells. Because of the rapid endocytosis and quenching of the fluorescein, these RHPAs must be analyzed within the first 15 min after the labeling for the ligand. This time constraint may prevent detection of secretion from target cells that secrete more slowly and require the full 3 to 4 hr for detection by RHPA. In other words, the dual-detection of ligand binding and secretion may not always be accurate. Some target cells may appear to be non-secretory. Other target cells may be secretory; however, they may have degraded the fluorescein-bound ligand and become undetectable. The problem with interpretation stems from the fact that one may not be able to tell if a secretory cell is secreting because of direct stimulation by the ligand, or because of a paracrine interaction from a target cell.

Techniques that employed avidin peroxidase detect the enzyme with diaminobenzidine.[2-5,9] This is made just before use by dissolving 4.5 mg of nickel ammonium sulfate and 7 to 10 mg of DAB in 0.05 M acetate buffer and then adding 15 μl of 30% hydrogen peroxide. The DAB is filtered through Whatman Grade I filter paper. Before adding the DAB, the phosphate buffer on the cells is removed by treating the cells with 0.05 acetate buffer (because the DAB precipitates with the phosphate ions). Then, the freshly made DAB is added for 6 to 7 min. It appears as a blue-black reaction product on or inside the labeled cells.

2. Tests of Specificity and Sensitivity of the Detection System

Once cytochemical labeling is established, the biotinylated ligands are subjected to a series of three tests. In the first, different concentrations (0 to 10 nM) are applied to the cells for 10 min to learn the point at which binding sites are saturated. This is quantified by counting the percentages of cell bound to the ligand. In most tests, no more cells are bound after 1 nM of the biotinylated ligand. An example is shown in Figure 2. Nearly 27% of pituitary cells were bound by 1 to 10 nM of biotinylated GHRH after exposure for 10 min. This matches the known percentages of growth hormone cells in the population; however, dual labeling is needed to prove that these are growth hormone (GH) bearing cells. These tests show that labeling can be detected following the use of physiological concentrations of the biotinylated ligand.

The specificity of the labeling is tested by competing the biotinylated ligand with its non-biotinylated counterpart, or related peptides. One of these tests is also

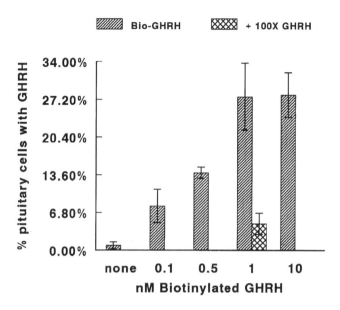

FIGURE 2

Biotinylated GHRH was added for 10-min to 2-day cultures of pituitary cells from proestrous female rats. After fixation, the biotinylated ligand was detected by avidin peroxidase. Counts of labeled cells showed that maximal percentages were reached following exposure to 1 nM of the ligand. If 100 nM GHRH were added to compete with the biotinylated GHRH for binding sites, the second graph shows that the percentages of labeled cells dropped to background levels.

shown Figure 2. When 100× excess (100 nM) GHRH was added to 1 nM biotinylated GHRH during the 10 min incubation, the percentages of labeled cells had dropped to less than 2%. Thus, the GHRH competed effectively for the binding sites. Parallel experiments are run in which other ligands such as CRH or GnRH are used to compete with the biotinylated ligand. Other, non-biotinylated ligands should have no effect on binding unless they bind to the same receptor or potentiate binding by the ligand in question.[26]

The third series of tests determines the optimal time for the binding experiments. In this group, a saturating concentration of biotinylated ligand is added for different times ranging from 30 sec to 30 min at 37°C. This saturating concentration is that which produces maximal percentages of labeled cells. The labeling intensity on the population is then quantified with time of exposure. In addition, the pattern of labeling is noted because it may change as the receptor-ligand complex is internalized. This has been described at the electron microscopic level in References 8, 9, 16, and 20. It will be briefly reviewed in the following paragraphs.

3. Changes in Distribution of Labeling with Time after Exposure

In most pituitary cell types, the labeling is initially (30 sec to 1 min) light and diffuse at the cell periphery. Electron microscopic studies show that it is mostly on the cell membrane. If the biotinylated ligand is exposed to the cells for longer periods at 2 to 4°C, the label will accumulate on the membrane. However, after warming to 37°C, internalization begins.

Within 3 min after warming, the labeling is in patches on the membrane and in small vesicles. One exception is the intermediate lobe cells. Labeling for biotinylated CRH is found in small vesicles within 30 sec to 1 min after exposure.[20] Internalization does not appear to require patching in these cells.

During the first 5 min, most target cells studied in this laboratory are labeled, although maximal labeling with biotinylated arginine vasopressin (AVP) requires 10 min.[15,26] After 10 min, the labeling is usually in larger vacuoles (endosomes), and by 30 min it may be in several compartments including the Golgi complex, lysosomes, and secretory granules. At this point, there is no proof that the labeling itself represents intact peptide because the lytic enzymes in the lysosomes may have digested the ligand.

Because of the rapid digestion of the ligand-receptor complex, the label in the cells disappears and the percentages of labeled target cells may actually decline after 15 min. Therefore, the purpose of the test of optimal times is to determine when maximal percentages of target cells are labeled. This dictates the timing of the experiments with different regulatory hormones. The objective is to use times that allow maximal labeling. As stated earlier, in the case of most ligands tested, the optimal time of exposure is 3 to 5 min. However, some ligands may require as much as 10 min.[12] These times are then used when any test of a regulatory factor is introduced.

D. Further Identification of the Labeled Target Cell

Once cytochemical labeling is established, there are a number of experiments one can do to learn more about the regulation of the expression of the receptor, binding to the ligand, and the types of target cells. First, one can run dual immunocytochemistry experiments to identify the target cell on the basis of its hormone content. This was first done in our laboratory in 1983 and it represents a standard method of identifying target cells today.[3,22] The approach involves the use of immunolabeling detection methods that contrast with those used for the detection of the ligand. We usually use immunoperoxidase methods with a contrasting colored substrate such as orange diaminobenzidine or alpha amino carbazol.[3-5] However, one can use immunofluorescein (for paraformaldehyde fixed tissues only), or immunogold reactions that are not intensified by silver (a light pink label is evident). At the electron microscopic level, one can use peroxidase to detect the biotinylated ligand followed by immunogold labeling to detect antigens in the cells.[6,26]

Routinely, we first detect the biotinylated ligand with the nickel-intensified diaminobenzidine as described above. It gives a blue-black reaction product. Then, the immunolabeling is run with DAKO rapid streptavidin peroxidase kits as described in the kit instructions.[22] After a 15 min blocking reaction in 1 to 3% bovine serum albumin, 5 to 10% normal goat serum, 1 to 2% non-fat dry milk, or other type of protein solution (depending on the type of antibody used) we expose the cells to the primary antisera usually diluted 1:10,000 to 1:50,000. The diluent is phosphate buffer (0.05 *M*) containing 2.5% crystalline human serum albumin and 1% non-fat dry milk (Carnation or generic brand). Usually only 30 min incubation in the primary antibody is needed, although 2 or 12 hr may produce better results. Then, the secondary antibody and detection system are applied for the times and temperature as described in the DAKO kit. The phosphate buffer is then washed out of the cells with 0.05 *M* Tris buffer used to make the orange DAB. Just before use, 7 to 10 mg of DAB are dissolved in 50 ml of 0.05 *M* Tris buffer, pH 7.6. Then, 25 μl of 30% hydrogen peroxide is added and the solution is mixed and filtered with Whatman Grade I filter paper. The solution is added to the cells for 6 to 7 min and the cells are then washed in Tris buffer. The reaction product should be orange to amber in the cells.

A number of immunolabeling controls must be done to prove that the two sequential labeling reactions have not interacted or interfered with one another. First, method controls that remove the primary antibody from the sequence should result in no amber labeling in the dual-labeled fields. Similarly, as the primary antibody is diluted, the amber labeling for that antigen should decrease. Third, the primary antibody should be absorbed with its antigen for 2 to 12 hr and then used in the labeling protocol. Absorption with excess antigen (100 fold) should result in no amber labeling, or a reduction in labeling as more antigens are added. Collectively these controls show that the antisera and their detection systems are not interacting with the components from the first labeling protocol.

A fourth type of control is to run parallel sets of cells in single labeling protocols and analyze the fields for the percentages of single labeled cells. This will show baseline percentages that will be compared with those obtained after dual labeling. These cell counts will determine if the dual labeling protocol affected the detection of either the ligand or the antigens. Counts of each type of labeled cell after single labeling protocols should be comparable to counts of total cells labeled for either the ligand or the antigen in the dual protocols.

We have developed Lotus 1,2,3 or Excel spread sheets with formulas that automatically calculate the percentages used in these tests (as described in References 22 or 27). They can be compared immediately to the known percentages from the single labeled fields, and the dual protocol can thus be validated. We do not consider a dual-labeling protocol valid unless it produces the same numbers of labeled cells found in the single labeling protocols.

Usually problems with interference are seen as a reduction in the percentages of antigen-bearing cells because of wash-out of antigens during the first protocol. However, the second protocol may wash-out the label from the first detection system, especially if it is prolonged or run at higher temperatures. If there is interference or washout, there will be a reduction in the percentages of target cells, or antigen-bearing cells, or both. This may require the following adjustments. If fewer antigen bearing cells are found, then the concentration of the primary antiserum is increased. If the second labeling protocol decreases the percentages of cells with ligand binding, then we shorten the time in the primary antiserum found in the second protocol, and/or reduce the incubation temperature. We could also add a glutaraldehyde fixation step to help fix the DAB label before the second protocol.

E. Experiments with the Biotinylated Ligands and their Target Cells

Over the past 14 years, we have experimented with the effects of steroids and regulatory peptides on expression of receptors.[5,9,10,12,14,15,18] We have also seen differential labeling of individual target cells with the physiological state.[8,12,19,22,26] These experiments are facilitated by the fact that physiological conditions are used for the exposure to the ligand. Thus, the cells are not down-regulated. The cells remain responsive to other hormones and steroids, and secretion can be detected from individual target cells by reverse hemolytic plaque assays.[10] In the following sections, we will show the types of experiments that can be done to learn more about the factors that affect binding and expression of receptivity in the pituitary population.

1. Identification of Target Cells: Unusual Findings

Recent studies of binding by biotinylated Arginine vasopressin (AVP), GnRH, or GHRH have identified rather unusual groups of target cells.[10,15,22] As expected, AVP-bound cells contained and secreted adrenocorticotropin (ACTH) antigens.[10,15] However, AVP is also a secretagogue for thyroid stimulating hormone (TSH). In subsequent studies of target cells, 60% of AVP-bound cells contained TSH antigens and a similar percentage contained ACTH antigens.[15] Since these percentages added to over 110% of AVP bound cells, we concluded that some of the target cells stored both TSH and ACTH.[15] These findings then stimulated dual-immunolabeling experiments which confirmed the existence of these multipotential corticothyrotrophs.[15] They would not have been discovered had it not been for the work with the biotinylated AVP ligand. The discovery allowed us to postulate that AVP had a unique group of target cells that could support both the thyroid and the adrenal. We hypothesized that these cells might be very useful in cases where multihormonal support was needed, such as cold stress.[15]

More recently, dual labeling studies showed that most of the GnRH bound target cells contained gonadotropin antigens, as expected.[22] Furthermore, their percentages increased appropriately as the population of gonadotrophs was prepared for ovulatory surge activity. However, our recent studies of gonadotropin mRNA expression also showed that during diestrus and proestrus (just before ovulation in the rat), cells with growth hormone antigens expressed gonadotropin mRNAs.[27] They appeared to augment the population of gonadotrophs during the proestrous preovulatory period. To further test their biological significance, biotinylated GnRH was detected in dual-labeling protocols with GH.[22] Nearly 40% of GH cells expressed receptors for the biotinylated analog by the afternoon of diestrus, and this expression persisted through the morning of proestrous. Then there was a rapid decline in expression after proestrous. This suggested that the multipotential somatogonadotrope is a transitory cell type designed to support the gonadotropin surge activity. Part of the support involves the expression of GnRH receptors. Figure 3 illustrates a somatotrope detected by gray label bound to biotinylated GnRH detected by black label.

These data led to the next series of tests of biotinylated growth hormone releasing hormone. The rationale was as follows. If the new GnRH target cells were

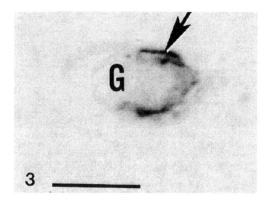

FIGURE 3
Growth hormone cell showing dense blue-black label for biotinylated GnRH (arrow) and gray (amber) label for growth hormone. Bar = 5 μm.

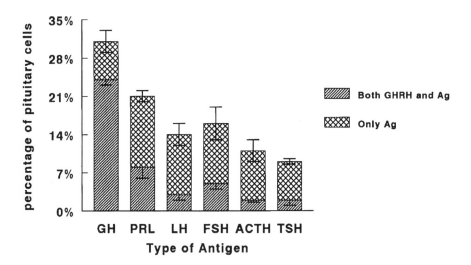

FIGURE 4
Dual label for GHRH and each of the pituitary hormones in a 2-day culture of cells from male rat pituitaries. The percentages of each cell type are indicated by the main bars. The subdivisions show the extent to which each cell type binds GHRH. In the male rat, GHRH is bound by cells with GH and prolactin antigens. Very few of the other cell types (less than 3% of the total population) bind GHRH.

truly multipotential, then we would expect to see binding of biotinylated GHRH by hormone bearing gonadotrophs during proestrus. Figure 4 shows that most GH cells and some prolactin cells bind GHRH in the male. Very few of the other cell types bind the hormone. Yet, in proestrous females (Figure 5), there is a proportionate increase in the percentages of cells with LH and FSH antigens that bind GHRH. This supports the hypothesis that functional somatotrophs actually become transitory gonadotrophs.

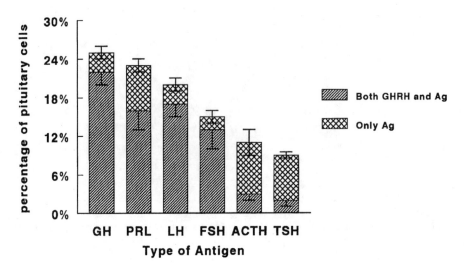

FIGURE 5

Dual label for GHRH and each of the pituitary hormones in a 2-day culture of cells from proestrous female rats. The percentages of each cell type are indicated by the main bars. The subdivisions show the extent to which each cell type binds GHRH. In the proestrous female rat, some of the GH cells produce gonadotropins.[27] Thus, the GHRH binds cells with GH, prolactin *and gonadotropin* (LH and FSH) antigens. Compare the results in this graph with those in Figure 4.

Thus, the dual-labeling experiments can often produce surprising data about the target cell populations. The data have raised questions about previous assumptions that pituitary cells produce only one type of hormone and cannot support other cell types. Not only are there multihormonal cells in the pituitary (producing combinations of pituitary hormones), there are also multifunctional cells that produce multiple types of receptors.

2. Regulatory Factors that Stimulate Binding or Receptivity to Ligands

In past studies, we have shown that regulatory hormones can affect binding and actually change the percentages of target cells as well as the intensity of the labeling. The labeled cells are counted and the values are expressed as a percentage of total target cells in the population. Several examples will be noted in this report. First, specific regulatory peptides may promote binding.[9,10,14,15] In experiments with CRH and arginine vasopressin, the cells were first pretreated with CRH or AVP for 1 hr before they were exposed to the biotinylated ligands. Pretreatment with AVP increased the percentages of corticotrophs that bind biotinylated CRH. Similarly, CRH pretreatment increased the percentages of corticotrophs that bind biotinylated AVP. Thus, certain regulatory peptides may have a potentiating effect on binding. This is not likely due to mitogenic effects, because the time is too short. Rather it may be due to the fact that the two regulatory peptides work via separate sets of

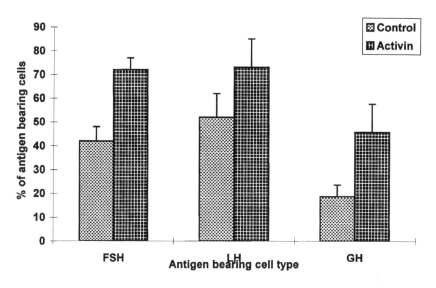

FIGURE 6

Activin regulates expression of GnRH receptors in pituitary cells in the cycling female rat. There is an increase in the percentage of each cell type that binds biotinylated GnRH. This includes the multipotential cell with growth hormone (GH) antigens.

second messengers on the same sets of cell. CRH acts via adenylate cyclase and AVP works through calcium and protein Kinase C. In our studies of second messenger activity, we showed that activation of these last 2 second messengers facilitates binding to CRH.[14]

Another regulatory peptide that promotes binding to gonadotropin releasing hormone is activin. Activin is known to promote the synthesis of GnRH receptor mRNA and therefore may work by enhancing expression.[28-31] After an overnight exposure to activin, cells from diestrous rats increased their expression of GnRH binding to levels similar to those found in proestrous rats. This suggests that activin may be a mediator in the normal increase in GnRH target cells during diestrus. Figure 6 shows that activin increases binding by LH and FSH gonadotrophs.

Activin also appears to have specific effects on the population of growth hormone cells that bind biotinylated GnRH. It stimulates more GH cells to bind the GnRH (Figure 6) in cells from diestrous rats. Thus, activin's effects may be to modulate the transition from a growth hormone function to a gonadotrope function. In this light, it is important to note that activin inhibits GH secretion and synthesis while stimulating that of gonadotropins, particularly FSH.[32-34]

3. Regulatory Factors that Inhibit Binding or Receptivity to Ligand

Factors that inhibit binding to corticotrophs by biotinylated CRH include glucocorticoids[9,18] and calcium or sodium channel blockers.[10,14] Similarly, testosterone inhibits biotinylated GnRH binding to gonadotrophs.[5] The steroid inhibition may reflect negative feedback effects, including inhibition of receptor synthesis. The rapid effects by ion channel blockers have not been easy to explain.[10,14]

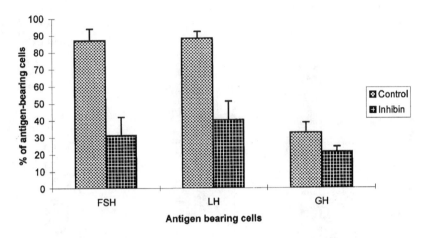

FIGURE 7

Inhibin regulates expression of GnRH receptors in pituitary cells in the cycling female rat. There is a reduction of GnRH binding to cells with LH or FSH antigens (gonadotrophs). No significant reduction is seen among cells with GH antigens, however.

A recent study shows that inhibin, which inhibits follicle stimulating hormone synthesis and release also causes a decrease in binding by biotinylated GnRH. The data show dramatic (70%) losses in the percentages of FSH cells that bind biotinylated GnRH (Figure 7). This peptide may thus regulate gonadotrophs partly through their expression of receptors for biotinylated GnRH.[35]

4. Quantitative Analysis of Binding by Densitometric Measurements

During the past 11 years, we have also quantified binding to individual cells either by counting the number of gold particles, or number of labeled sites[8,9] or by assaying labeling density.[14,16,18,19,20,26,35] Both approaches provide a semi-quantitative estimate of the amount of label per cell and its distribution with respect to time.[8,9,26] These approaches can also be used to test the effects of regulatory hormones on binding capacity of individual cells.[26,35]

The analysis is done with a Bioquant MEG IV image analysis system which detects labeling by both color and density. After the labeling threshold is established and appropriate background corrections are made, the integrated optical density of the label can be read automatically. Figure 8 shows an example of the results of density measurements. It shows a histogram of labeling densities measured for 100 GnRH target cells per experimental group.[35] The cells were being tested for GnRH labeling after inhibin treatment. The shift to a population of cells with a lower labeling density (seen as higher numbers in this graph) is evident.

Similarly, activin enhances both the area and the density of labeling for biotinylated GnRH on individual cells. Figure 9 illustrates an example of the types of measurements obtained. It shows the activin-mediated increased area of GnRH label per cell. A density histogram was also obtained for these cells. The shift in

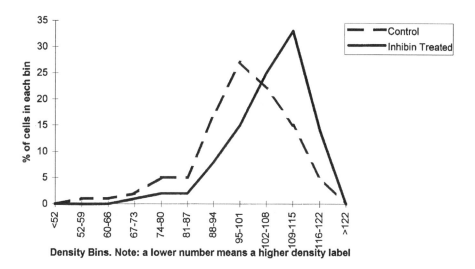

FIGURE 8

Densitometric analyses of the label for Biotinylated GnRH (Bio-GnRH) on individual target cells shows that inhibin causes a shift in labeling to a population of target cells with weak label. This histogram shows this shift. The higher numbers shown by the histogram for the inhibin treated cells represent lower density labeling (more light is being transmitted by the label).

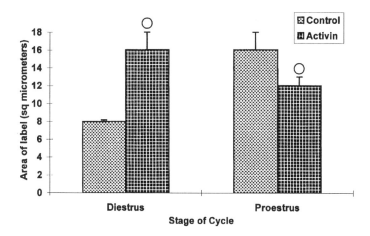

FIGURE 9

Another type of analysis includes the area of label for biotinylated GnRH. This graph shows the activin-stimulated increase in the area of label for the ligand on individual target cells. The circle over the bar graph indicates a significant increase in label area seen in the diestrous rat population. In proestrous rats, there is a significant decrease in the area of the label for biotinylated GnRH.

the population is towards more GnRH target cells with higher density label (data not shown).

F. Summary and Conclusions

The purpose of these studies was to describe current methods in this laboratory that are used for the detection of biotinylated ligands attached to pituitary target cells. The primary function of this work has been to identify target cells and learn the extent to which the receptor population is expressed as the physiological state is altered. We have also focused on effects of regulatory factors on receptor expression. A surprising outcome has been the discovery, in several cases, of multifunctional cells that express unusual combinations of hormones. The discoveries have stimulated hypotheses that pituitary cells may serve "cocktails" of hormones when needed. For example, since cold stress stimulates both ACTH and TSH secretion, perhaps the AVP target cell that produces both hormones is most effective in the response to this type of stress. In another example, since growth hormone may be used in the normal course of reproduction, perhaps the transitional somatogonadotrophs serve a "cocktail" of GH and gonadotropins to promote ovulation.[27] GH secretion is attenuated by activin in these cells[31-34] so that the transition to a gonadotroph can be effected. These discoveries would not have been possible without the quantitative affinity cytochemistry tools used during the past 14 years.

This report has also shown additional approaches to a study of regulation of expression of receptors by different subsets of target cells. We can detect inhibition or stimulation of receptor expression both by changes in numbers of target cells as well as by changes in the density of labeling on each target cell. Specific changes on each cell can be interpreted as changes in the rate of endocytosis, degradation, or the number of binding sites available to the biotinylated ligand. At this point, the technology does not allow one to differentiate between these possibilities. Nevertheless, the data suggest that changes in expression by individual cells can be regulated by specific modulatory hormones or their second messengers in order to affect their sensitivity to secretagogues.

References

1. Bayer, E.A., Skutelsky E., and Wilchek, M., The avidin-biotin complex in affinity cytochemistry. *Methods Enzymol.*, 62, 308–315, 1979.
2. Childs, G.V., Naor, Z., Hazum, E., Tibolt, R., Westlund, K.M., and Hancock, M.B., Localization of biotinylated gonadotropin releasing hormone on pituitary monolayer cells with avidin-biotin peroxidase complexes. *J. Histochem. Cytochem.*, 31, 1422–1425, 1983.
3. Childs, G.V., Naor, Z., Hazum, E., Tibolt, R., Westlund, K.N., and Hancock, M.B., Cytochemical characterization of pituitary target cells for biotinylated gonadotropin releasing hormone. *Peptides*, 4(4), 549–555, 1983.

4. Westlund, K.N., Wynn, P.J., Chmielowiec, S., Collins, T.J., and Childs, G.V., Characterization of a potent biotin-conjugated CRF analog and the response of anterior pituitary corticotrophs. *Peptides,* 5, 627–634, 1984.

5. Tibolt, R.E. and Childs, G.V., Cytochemical and cytophysiological studies of GnRH target cells in the male rat pituitary: Differential effects of androgens and corticosterone on GnRH binding and gonadotropin release. *Endocrinology,* 117(1), 396–404, 1985.

6. Naor, Z. and Childs, G.V., Action and binding of gonadotropin releasing hormone to target cells in pituitary and gonads. *Intl. Rev. Cytology,* 103, 147–187, 1986.

7. Niendorf, A., Dietel, M., Arps, H., Lloyd, J., and Childs, G.V., Visualization of binding sites for parathyroid hormone (1-84) on cultured kidney cells with Biotinyl-β-PTH (1-84). *J. Histochem. Cytochem.,* 34, 357–361, 1986.

8. Childs, G.V., Hazum, E., Amsterdam, A., Limor, R., and Naor, Z., Cytochemical evidence for different routes of GnRH processing by large gonadotrophs and granulosa cells. *Endocrinology,* 119, 1329–1338, 1986.

9. Childs, G.V., Morell, J.L., and Aguilera, G., Cytochemical studies of CRF receptors in anterior lobe corticotrophs: Binding, glucocorticoid regulation and endocytosis of [Biotinyl-Ser[1]] CRF. *Endocrinology,* 119, 2129–2142, 1986.

10. Childs, G.V., Unabia, G., Burke, J.A., and Marchetti, C., Secretion from corticotrophs after avidin-fluorescein stains for biotinylated ligands (CRF or AVP). *Am. J. Physiol.,* 252, (Endocrinol. Metab. 15), E347–E356, 1987.

11. Marchetti, C., Childs, G.V., and Brown, A.M., Membrane currents of identified isolated rat corticotrophs and gonadotrophs. *Am. J. Physiol.,* 252, (Endocrinol. Metab. 15), E340–346, 1987.

12. Lloyd, J.M. and Childs, G.V., Changes in the number of GnRH-receptive cells during the rat estrous cycle: biphasic effects of estradiol. *Neuroendocrinology,* 48, 138–146, 1988.

13. Niendorf, A., Dietel, M., Arps, H., and Childs, G.V., A novel method to demonstrate parathyroid hormone binding on unfixed living target cells in culture. *J. Histochem. Cytochem.,* 36, 307–309, 1988.

14. Childs, G.V. and Unabia, G., Activation of protein Kinase C and voltage dependent calcium channels enhances binding of CRH by anterior pituitary cells. *Mol. Endo.,* 3, 117–126, 1989.

15. Childs, G. V., Westlund, K., and Unabia, G., Characterization of anterior pituitary target cells for arginine vasopressin: including cells that store adrenocorticotropin, thyrotropin-β and both hormones. *Endocrinology,* 125, 554–559, 1989.

16. Childs, G. V., Yamauchi, K., and Unabia, G., Localization and quantification of hormones, ligands and mRNA with affinity-gold probes. *Am. J. Anat.,* 185, 223–235, 1989.

17. Childs, G.V., Lloyd, J., Unabia, G., and Rougeau, D., Growth and secretory responses of enriched populations of corticotrophs. *Endocrinology,* 125, 2540–2549, 1989.

18. Childs, G.V. and Unabia, G., Rapid corticosterone inhibition of CRH binding and ACTH release by enriched populations of corticotrophs: Counteractions by AVP and its second messengers. *Endocrinology,* 126, 1967–1975, 1990.

19. Sasaki, F., Wu, P., Rougeau, D., Unabia, G., and Childs, G.V., Cytochemical studies of responses of corticotrophs and thyrotrophs to cold and novel environment stress. *Endocrinology,* 127, 285–297, 1990.

20. Childs, G.V., Subsets of pituitary intermediate lobe cells bind CRH and secrete ACTH/CLIP in a reverse hemolytic plaque assay. *Peptides,* 11, 729–736, 1990.

21. Vigh, S., Arimura A., Gottschall, P.E., Kitada, C., Somogyvari-Vigh A., and Childs, G.V., Cytochemical characterization of anterior pituitary target cells for the neuropeptide, pituitary adenylate cyclase activating polypeptide (PACAP), using biotinylated ligands. *Peptides*, 14, 59–65, 1993.

22. Childs, G.V., Unabia, G., and Miller, B.T., Cytochemical detection of GnRHbinding sites on rat pituitary cells with LH, FSH and GH antigens during diestrous upregulation. *Endocrinology*, 134, 1943–1951, 1994.

23. Hsu S-M., Raine L., and Fanger H., Use of avidin-biotin peroxidase complex (ABC) in immunoperoxidase techniques. A comparison between ABC and unlabeled antibody (PAP) procedures. *J. Histochem. Cytochem.*, 29, 577–580, 1981.

24. Smith, J.S., Miller, B.T., Knock, S.L., and Kurosky, A., Biotinylated peptides/proteins: I. Determination of stoichiometry of derivatization. *Anal. Biochem.*, 197, 247–253, 1991

25. Miller, B.T., Collins, T.J., Nagle, G.T., and Kurosky, A., The occurrence of O-acylation during biotinylation of gonadotropin releasing hormone and analogs: evidence for a reactive serine. *J. Biol. Chem.*, 267, 5060–5069, 1992.

26. Childs, G.V., Westlund High, K.N., Tibolt R.E., and Lloyd J.M., Hypothalamic regulatory peptides and their receptors: Cytochemical studies of their role in regulation at the adenohypophyseal level. *J. Elec. Mic. Tech.*, 19, 21–41 1991.

27. Childs, G.V., Unabia, G., and Rougeau, D., Cells that Express Luteinizing Hormone (LH) and Follicle Stimulating Hormone (FSH) Beta (β) Subunit mRNAs during the Estrous Cycle: The major contributors contain LHβ, FSHβ and/or Growth Hormone, *Endocrinology*, 134, 990–998 1994.

28. Braden T.D. and Conn, P.M., Activin A stimulates the synthesis of gonadotropin-releasing hormone receptors. *Endocrinology*, 130, 2101–2105, 1990.

29. Braden, T.D. and Conn, P.M., The 1990 James A.F. Stevenson Memorial Lecture. Gonadotropin-releasing hormone and its actions. *Can. J. Physiol. Pharmacol.*, 69, 445–458, 1990.

30. Conn, P.M., Janovic J.A., Stanislaus, D., Kuphal D., and Jennes L., Molecular and cellular bases of gonadotropin releasing hormone action in the pituitary and central nervous system. *Vitamins and Hormones*, 50, 152–201, 1995

31. Fernandez-Vazquez, G., Albarrancin, C.T., Kaiser, U.B., and Chin, W.W., Transcriptional activation of the gonadotropin-releasing hormone receptor (GnRHR) gene by activin A in alpha T3-1 cells. *Proc. Mtgs. Endocrine Soc.*, P2–36, 1995.

32. Bilezikjian, L.M., Corrigan, A.Z., and Vale, W., Activin-A modulates growth hormone secretion from cultures of rat anterior pituitary cells. *Endocrinology*, 126, 2369–2376, 1990.

33. Billestrup N., Gonzalez-Manchon, C., Potter E., and Vale, W., Inhibition of somatotroph growth and growth hormone biosynthesis by activin *in vitro*. *Mol. Endocrinol.*, 4, 356–362, 1990.

34. Bilezikjian, L.M., Corrigan, A.X., and Vale, W.W., Activin-B, Inhibin-B and Follistatin as autocrine/paracrine factors of the rat anterior pituitary. Int Symp on Inhibin and Inhibin-Related Proteins. *Challenges in Endocrinology and Modern Medicine*, H.G. Burger, Ed., 1–19, 1993.

35. Childs, G.V., Miller, B.T., and Miller, W.L., Differential Effects of Inhibin on Gonadotropin Stores and Gonadotropin Releasing Hormone Binding to Pituitary Cells from Cycling Female Rats. *Endocrinology*, 138, 1577, 1997.

Chapter **4**

Steroid Hormone Receptors and the Assessment of Feedback Sensitivity

Robert J. Handa, Richard H. Price Jr., and Melinda E. Wilson

Contents

0-8493-3363-6/98/$0.00+$.50

I. Introduction

The concept of feedback regulation of hormone secretion represents a cornerstone upon which much of neuroendocrine research has been built. Some of the best known examples of hormone feedback involve the regulation of hypothalamic and anterior pituitary function by steroid hormones. Steroid hormones, secreted from the adrenal gland or gonads in response to circulating pituitary hormones, interact with neuroendocrine tissues to ultimately inhibit or augment their own secretion. This long loop feedback system is an integral feature of the hypothalamo-pituitary-adrenal (HPA) and hypothalamo-pituitary-gonadal (HPG) axes and represents an efficient way in which the central nervous system can monitor end organ secretion and subsequently regulate anterior pituitary function in order to maintain homeostasis or regulate reproductive function.

The effects of steroid hormones on the brain are widespread. Steroid hormones have been shown to modulate a variety of behaviors and functions (for review see References 1 and 2), only some of which concern the central regulation of neuroendocrine function. Consequently, a requirement in the development of the concept of feedback regulation by steroid hormones is the demonstration that specific receptors for these hormones exist in the brain areas known to be involved in the central regulation of anterior pituitary function.

II. Mechanisms of Steroid Hormone Action

A. Intracelllular Receptors Mediating the Genomic Effects of Steroid Hormones

The classical mechanism by which steroid hormones affect cellular function involves the binding to specific intracellular receptors which function to enhance or repress gene transcription. The receptors for steroid hormones belong to a single superfamily of proteins which include receptors for androgen, estrogen, progesterone, glucocorticoid and mineralocorticoids as well as thyroid hormone, retinoic acid, vitamin D, and a host of related proteins termed "orphan receptors".[3] In the presence of their cognate ligand, steroid hormone receptors are rapidly transformed from an inactive to an active state. This transformation involves the removal of some receptor associated proteins such as heat shock protein,[4] conformational changes, and homodimerization. As a result, the receptor can bind to specific target sites on DNA (termed hormone response elements) with high affinity. Finally, the binding of receptor dimers to hormone response elements in the promoter region of steroid sensitive target genes acts to enhance or repress transcription.

Target gene specificity for steroid hormone receptors is conferred by receptor specific hormone response elements on DNA. However, steroid hormone receptors are promiscuous. For example, it appears that glucocorticoids, mineralocorticoids, progesterones, and androgen work through a common, or very closely related hormone response element on DNA.[5] Thus, target gene specificity is also a function of cell-specific regulation of receptor levels, tissue specific metabolism of hormone,[6] or the presence of other nuclear proteins which may interact with the receptor and dictate specificity or even the direction of regulation (enhancement vs. repression).[7] The concept of "complex response elements" upstream of hormone responsive genes, in which multiple transcription factors interact to influence the actions of a steroid hormone receptor is currently the focus of intense investigation.

B. Non-Genomic Effects of Steroid Hormones

Not all of the effects of steroid hormones can be explained by the relatively slow mechanisms associated with changes in gene expression following steroid hormone receptor activation. For example, it is well known that ACTH secretion can be inhibited within minutes of changes in plasma corticosterone levels.[8] The rapid action of corticosterone cannot be blocked by cyclohexamide, suggesting that *de novo* protein synthesis is not involved. These rapid actions of steroid hormones suggest a cell membrane site of action.[9]

Evidence for membrane-mediated responses comes from electrophysiological studies demonstrating that in some tissues, changes in membrane potentials can be found immediately after glucocorticoid[10] or estrogen[11,12] administration. Recent studies have also shown that somes metabolites of progesterone can bind to membrane receptors utilized by other neurotransmitter systems. For example, it is now well known that allopregnanolone or $3\alpha,21$-dihydroxy-5α-prengnan-20-one, can bind to $GABA_A$ receptors, and by doing so will inhibit postsynaptic membrane responses.[13]

Although these studies demonstrate the presence of rapid membrane effects consistent with fast negative feedback on hormone secretory axes, the biochemical evidence for the existence of specific membrane receptors for steroid hormones remains sparse. Towle and Sze[14] were able to demonstrate a relatively high affinity receptor for estrogen and progesterone in purified membranes of rat brain. More recently, Orchinik et al.,[15] were able to use conventional binding assays to demonstrate a high affinity receptor for corticosterone in membrane preparations of the newt brain.

The non-genomic mechanisms through which steroids act to influence hormone secretion remain elusive. Consequently, the remainder of this chapter will focus on methods used to examine steroid action mediated through the classical intracellular receptor. A more thorough examination of the rapid membrane effects of steroid hormones can be found in a recent review by Ramirez et al.[16]

III. Do Levels of Intracellular Steroid Hormone Receptors Reflect Feedback Sensitivity?

A recurrent theme underlying many studies examining steroid hormone receptors in neuroendocrine tissues is that the concentration of receptor in a given population of cells may reflect the sensitivity of those cells to circulating steroid hormone. On an absolute basis this is probably true; the presence of receptor indicates cells that are sensitive to the hormone, whereas the absence of receptor indicates that the cell is not sensitive. However, when examining more subtle changes in receptor levels, the assumption that receptor number is equivalent to sensitivity is perhaps an oversimplification. It is not difficult to find reports of small changes or differences in receptor levels which do not necessarily correlate with the physiological observations. For example, it is well known that the gonadotropin positive feedback response to estrogen and the female typical behavioral response to estrogen, are absent or low in male rats.[17] This can be interpreted as representing a decreased sensitivity to estrogen by males. The comparison of estrogen receptor levels in hypothalamic nuclei involved in regulating positive feedback responses and reproductive behaviors shows the presence of a sex difference (males < females), however, the magnitude of this sex difference (approximately 10 to 20%) is small,[18] particularly if one considers the magnitude of the difference in estrogen responsiveness between males and females.

Another example of apparent changes in feedback sensitivity which does not necessarily correlate with changes in receptor number comes from studies examining the negative feedback regulation of ACTH release following stress. Normally, glucocorticoid receptors are activated by stress levels of corticosterone and these receptors function to inhibit further ACTH release from the anterior pituitary gland. In the presence of estrogen, this feedback inhibition is impaired. However, there is no change in glucocorticoid or mineralocorticoid levels in the hypothalamus or hippocampus.[19]

These examples illustrate the complexity of neuroendocrine regulatory systems and the potential flaw in the simplistic assumption that changes in receptor number reflect corresponding changes in sensitivity. Alternatively, it could be argued that in any given tissue, only a few important cells are involved in feeback regulation and thus, it is necessary to focus on receptor levels in these identified cells. Both arguments are probably correct. Recent studies using molecular techniques and examining steroid hormone receptor action in peripheral tissues have repeatedly described multiple ways in which steroid hormone receptor action can be altered by other "non-receptor factors". A "composite hormone response element" has been described for glucocorticoid receptor regulation in which the presence of an upstream AP-1 site may dictate the degree and the direction in which glucocorticoid receptors can regulate the target gene.[7] Similar mechanisms have been described for estrogen and androgen receptors. Recently, Yeh and Chang[20] have described in prostate gland cells, the presence of a receptor associated protein ARA70 which, when present in the appropriate amounts, will increase the transcriptional response of a given number of androgen receptors by as much as 10 fold. A similar enhancement of transcriptional efficiency was described for the estrogen receptor by the CREB binding protein.[21] Unfortunately for neuroendocrinologists, the fact that steroid hormone receptors are generally expressed at low levels in brain tissues makes it difficult at best, to measure changes in these low levels of steroid hormone receptors in one phenotype out of a population of hundreds of different neuronal phenotypes.

With these caveats in mind, this chapter will examine some of the techniques by which it is possible to assess the tissue and cellular distribution, dynamics, synthesis and regulation of steroid hormone receptors in neuroendocrine tissues. Only by the use of multiple techniques to examine steroid hormone receptors can one make educated guesses as to the feedback sensitivity of neuroendocrine tissues to a given hormone.

IV. Methods for the Measurement of Steroid Hormone Receptors

Although steroid hormone receptors are found in brain, they are often found at low levels in comparison to peripheral tissues such as the prostate gland and uterus. This fact, coupled with the obvious complexity of brain tissue, indicates that careful evaluation of methodology is necessary prior to choosing one technique. Essential in this choice is understanding the advantages and disadvantages which each technique affords. The following discussion of methodological approaches will focus on the utility of each method in situations of cellular heterogeneity and low copy number which are characteristics of neuroendocrine tissues and steroid hormone receptors.

A. *In Vitro* Binding Assays

A considerable amount of information regarding the distribution, levels and dynamics of steroid hormone receptors can be obtained by the use of *in vitro* binding assays. *In vitro* binding assays generally involve the cellular fractionation of tissue by centrifugation techniques followed by the binding of cytosolic fractions or nuclear extract with a radiolabeled ligand. These types of protocols have been published previously.[18,19,22] Briefly, tissue is gently homogenized in Tris or phosphate buffers containing EDTA, molybdate, glycerol and dithiothreitol to stabilize the receptor. Centrifugation of the homogenate at low speed (1000 × g) will separate the cell debris and nuclei from the crude cytosolic fraction (supernatant). A further high speed (100,000 × g) centrifugation of the crude cytosol results in a relatively pure cytosolic preparation which can be used in binding assays. The pellet obtained from the initial low speed centrifugation is generally purified by centrifugation through a sucrose pad (0.32 M sucrose). The resulting pellets of purified nuclei are subjected to repeated exposures to high salt buffers (0.8 M KCl) in order to extract the DNA bound receptor. Finally, centrifugation separates the nuclear debris from the extracted nuclear protein which can be utilized in conventional binding assays.

The binding of radiolabeled ligand to unliganded receptor found in the cytosolic fractions is straightforward. Although the construction of saturation isotherms, by binding to increasing doses of ligand are preferred for quantitation, an estimate of receptor number can be made with a single concentration of ligand. This latter approach is preferred when receptors are being measured from small pieces of brain tissue such as those following microdissection. In both cases however, care must be taken in determining the exact length of time and the temperature of incubation. Each receptor has different requirements for optimum binding. For example, estrogen receptors show maximal binding characteristics when cytosols are incubated at 24°C for 2 to 4 hr. In contrast, androgen receptors show maximal binding when incubated at 4°C for 24 to 48 hr.

Binding of radiolabeled ligand to the occupied receptor in nuclear salt extracts involves the exchange of the radiolabeled ligand for the bound ligand. For some receptors, such as glucocorticoid and mineralocorticoid receptor, this exchange is inefficient or does not occur and thus, the measurement of occupied receptors cannot be accomplished. Estimates of occupied receptors must be made by measuring changes in unoccupied receptors in cytosolic fractions in the presence or absence of hormone.[23] Fortunately, for those measuring estrogen or androgen receptors, this exchange is rapid and complete and thus, estimates of occupied receptor can be easily determined.

The obvious advantage of using binding assays is their ease and rapidity. Data can be obtained within a day or two. Information regarding the amount of unoccupied (cytosolic) or occupied (nuclear) receptor can be obtained which allows one to make predictions regarding the amount of active receptor at a given time or following a given dose of hormone. Affinity, on-off characteristics, specificity and the number of binding sites can be easily obtained by the use of saturation curves, competition curves and time or temperature curves.

Disadvantages of *in vitro* binding assays are generally related to the fact that anatomical resolution is lost. Microdissection procedures allow the measurement of receptor in individual hypothalamic nuclei, however, this still consists of a quite heterogeneous population of cells. It is also difficult, using this technique to obtain an accurate estimate of total receptors (occupied and unoccupied) during conditions where hormone is present, since these parameters are not necessarily additive due to differences in tissue processing and losses during extraction.

Recent modifications of *in vitro* binding assays have been made which allow binding to occur on tissue sections.[24] Cytosolic receptor is measured following precipitation within tissue by the addition of protamine sulfate to the incubation buffer. Nuclear exchange is efficient for some receptors such as estrogen receptors. Following extensive washes to separate bound and free ligand, the tissue sections are exposed to radiation-sensitive film. The resulting image is easily quantifiable by standard densitometry. The advantages of this approach are obvious. Anatomical detail is maintained and, given the same caveats as above, the concentrations of occupied and unoccupied receptors can be delineated from adjacent tissue sections. Unfortunately, in the absence of high energy ligands, the use of this technique results in exposure times which could be considerable. Nonetheless, this approach represents a considerable advancement in the measurement of steroid receptor numbers in brain tissue.

B. Immunochemical Methods of Receptor Detection

The production of specific antibodies against steroid hormone receptors has allowed the anatomical mapping of receptors within the brain using standard immunocytochemical procedures, or the determination of receptor levels and isoforms using Western blot analysis. The immunocytochemical procedure results in exceptional anatomical detail. The ability to couple this technique with *in situ* hybridization, or with the immunocytochemical detection of other antigens provides a method by which to determine the phenotype of steroid hormone responsive cells.

For example, in the pituitary, FSH secretion and FSH mRNA expression are regulated by androgens.[25] Treatment of primary cultures of pituitary cells with androgen increases FSH-beta mRNA expression.[26] The probability of a direct effect of androgens on FSH containing gonadotrophs can be emphasized by the demonstration of androgen receptor immunoreactivity (AR-ir) in FSH expressing cells. Using dual labelling immunocytochemical procedures, AR-ir was detected in pituitary cells using nickel intensified diaminobenzidine (DAB) procedures. This results in a dark precipitate in the nuclear compartment of AR-ir cells. This procedure was followed by the immunocytochemical detection of FSH using DAB alone to yield a lighter oxidation product in the cytoplasm. Using this technique, androgen receptors can be visualized in the nucleus of FSH containing gonadotrophs (Figure 1). Cell counting analysis showed that over 90% of FSH containing cells were AR-ir positive. Examination of all AR-ir positive cells showed that 52% percent were FSH containing gonadotrophs, leading to the conclusion that androgen receptors are directly involved in regulating FSH synthesis and may, additionally, modulate the function of multiple cell types in the pituitary.

In contrast, the Western blot procedure, in which cell homogenates are electorphoresed through denaturing polyacrylamide gels prior to immunochemical staining, provides little anatomical detail, is generally less sensitive, but is probably easier to quantitate and can potentially demonstrate the size of the receptor protein and the presence of different isoforms. Few studies to date have used Western blot analysis for examination of steroid hormone receptors in brain tissue. This is probably due to the low levels of receptor present in neuroendocrine tissues, the lack of sensitivity of this technique and the extremely labile nature of steroid hormone receptors in the absence of tissue fixation.

As with any technique, there are some pitfalls which should be addressed. Any immunochemical technique, by its very nature, is specific for the protein to which the antibody is raised. Thus, it is possible that the antibody will not detect isoforms of the receptor protein, which would be detectable using binding assays. An example of this can be seen with immunocytochemical studies of the estrogen receptors. The recent elucidation of a second isoform of estrogen receptor (termed estrogen receptor-β) which shares limited homology with the original isoform of estrogen receptor (estrogen receptor-α)[27] has made earlier immunocytochemical studies of estrogen receptors incomplete. The discovery of alternate forms of estrogen receptor provides a convenient answer to the previously unresolved issue of discordant distribution of estrogen receptors when measured by *in vivo* autoradiography vs. immunocytochemistry.

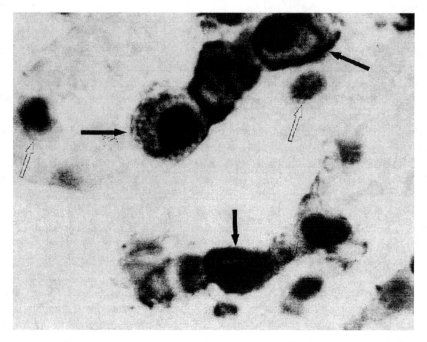

FIGURE 1

Photomicrograph showing the co-localization of androgen receptor immunoreactivity in follicle stimulating hormone immunoreactive cells within the anterior pituitary gland of the female rat. Androgen
receptor was detected using antibody PG-21 and nickel intensified diaminobenzidine. Its presence is
demonstrated by a darkly stained nucleus. FSH was detected using antibodies supplied by the National
Hormone and Pituitary Program and standard diaminobenzidine reaction. Its presence is demonstrated
by cytoplasmic staining. Filled arrows designate dual labeled cells, unfilled arrows designate single labeled
cells. (Magnification = 1000×)

Immunocytochemical techniques can also be criticized for the difficult nature
of quantitation. Because of the amplification steps required for double antibody
techniques, immunocytochemical data are semi-quantitative at best. Thus, the gain
of anatomical resolution offered by this technique is tempered by a loss in power
of quantification.

C. *In Vivo* Receptor Autoradiography

The earliest studies examining the distribution of steroid hormone receptors in brain
tissue used *in vivo* autoradiography for detection.[28] This technique provides excellent
anatomical detail and the ability to quantitate receptors in individual cells. By its
very nature, this technique measures all receptor forms with similar affinity for the
ligand. Briefly, animals are gonadectomized and/or adrenalectomized to remove any
endogenous source of steroid. Animals are then injected with large amounts of
radiolabeled ligand, sufficient to saturate the receptors. Animals are sacrificed and the
distribution of radiolabeled ligand is detected by sectioning the brain and opposing the

tissue to photographic emulsion coated cover slips. Non specific binding is determined in animals who receive excess competing steroid prior to the radiolabeled ligand.

Unfortunately, *in vivo* autoradiogaphical techniques are tedious and time consuming. It is not unusual when using tritiated ligands, for exposures to last 6 months to 2 years (or more!) in order to adequately detect binding. However, with the advent of high energy isotope labeled steroids, exposure times can be reduced considerably.

Another caveat is the inability to use *in vivo* autoradiography to measure steroid hormone receptors under conditions (physiological or otherwise) in which circulating steroid is present. The circulating steroid hormones compete with radiolabeled ligand for the binding site and can effectively reduce the signal to non-detectable levels. Thus, removal of the source of hormone (gonadectomy or adrenalectomy) is essential and limits the usefullness of this technique.

V. Methods for the Measurement of Receptor Gene Expression

An important step in the assessment of steroid hormone feedback is a knowledge of the regulation of steroid receptor gene expression. Information concerning the mechanisms by which steroid hormone receptor expression is regulated is necessary to our basic understanding of the factors which modulate neuroendocrine feedback. Additionally, factors which induce or prevent receptor gene expression could be considered as potential targets for pharmacological intervention in disease states.

With the cloning and characterization of the steroid hormone receptor genes,[29-31] basic molecular techniques can be used to measure steady state levels of receptor mRNA. Each technique has its advantages and disadvantages for the measurement of low levels of mRNA in brain tissues which is characteristic of steroid hormone receptor genes.

A. Northern Blot Hybridization Analysis

Northern blot hybridization analysis can be used to semi-quantitate amounts of a specific mRNA from a total RNA pool derived from a tissue homogenate. Typically, total RNA is size fractionated by electrophoresis in agarose gels, transferred to nylon membrane by capillary or electrophoretic action and specific mRNAs are detected by hybridization with labeled cDNA probes. Northern blot analysis is particularly useful for determining the molecular size of the mRNA of interest, or to determine if more than one species of mRNA exists. For example, using Northern blot analysis, we have determined that two forms of AR mRNA exist in brain with region specific distributions,[32] and that two forms of estrogen receptor alpha mRNA exist in pituitary but not brain.[33] The highest levels of the smaller form of AR mRNA are found in the cortex[34] whereas equivalent levels of both forms are found in the hypothalamus.[32]

This observation has led to further studies which defined the 5' untranslated region (UTR) of the small and large form of AR mRNA. It appears that the smaller form of AR mRNA lacks much of the 5'-UTR of the larger form.[35] Since the 5'-UTR is involved in translational regulation of the AR gene,[36] this suggests an alternate way by which levels of receptor can be altered in a tissue specific fashion.

The disadvantages of using Northern blot analysis are obvious. Northern blot analysis requires relatively large amounts of tissue as a source for total RNA. This limits its usefulness in brain research where anatomical resolution is required. This analysis is lacking in sensitivity and therefore the detection of steroid receptor mRNAs, which are generally found at low levels in brain tissue, is difficult for neuroendocrine studies using Northern blot analysis. Quantification of Northern blots presents an additional difficulty. Semi-quantification can be accomplished by densitometry of hybridized bands, but only following normalization to an internal control RNA. As a result of these potential difficulties, Northern blot analysis is not frequently used for examining the regulation of steroid hormone receptor mRNAs in neuroendocrine tissues.

B. *In Situ* Hybridization

In situ hybridization perhaps represents the best compromise between quantification and anatomical resolution. *In situ* hybridization utilizes radiolabeled or nonisotopically labeled cRNA or oligonucleotide probes which are hybridized directly to fresh frozen or fixed tissue sections. Following extensive washing and/or treatment with RNAse to remove nonspecific hybridization, the signal is visualized by exposure to X-ray film or by coating the slide mounted sections with photographic emulsion. Generally, film autoradiography offers the advantage of short exposure times, but with a loss of anatomical resolution. Emulsion coated slide autoradiography requires approximately 3- to 4-fold longer exposure times, but mRNA expression can be determined in individual cells by grain counting techniques. Recent advancements using nonisotopically labeled probes followed by immunochemical detection methods offers advantages of rapid development time, but with a loss of both sensitivity and the ability to quantify beyond cell counting methods. Several protocols have been published recently showing the distribution and regulation of steroid hormone receptor mRNAs in brain tissues.[37,38] The biggest advantages of the *in situ* hybridization technique are sensitivity, anatomical resolution, the ability to semi-quantify expression on an individual cell basis (using emulsion coated slide autoradiography), or within brain nuclei (using film autoradiography), and the ability to couple this technique with others such as immunocytochemistry[39] or reverse hemolytic plaque assays.[40] This allows the determination of steroid receptor gene expression in identified neuroendocrine cells.

An example of data which can be generated using this technique is illustrated in Figure 2. Using *in situ* hybridization and film autoradiography, progesterone receptor mRNA levels were determined in the medial preoptic area of 3-month-old, 8-month-old and 15-month-old female rats following ovariectomy and estrogen treatment (10 µg/kg BW, daily for 4 days). The construction of the cRNA probe

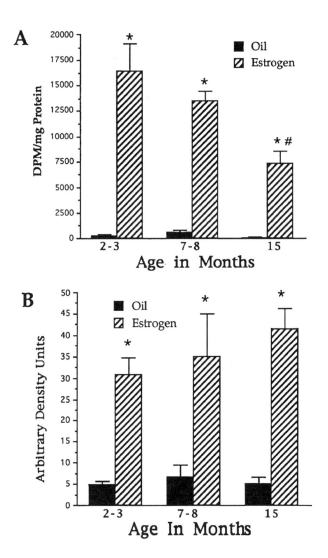

FIGURE 2

Induction of progesterone receptor mRNA levels by estrogen in the MPOA (A) and Arcuate n. (B) of young (3 months of age); middle-aged cycling (8 months of age), and middle aged noncycling (15 to 18 months of age) female rats. All animals were gonadectomized for 1 week prior to daily treatment with estradiol benzoate (10 μg/mg BW × 4 days). Progesterone receptor mRNA was detected by *in situ* hybridization using a probe which detects both the A and B form of receptor mRNA. Each bar represents the mean ± SEM of 4 to 6 determinations; * = significantly elevated vs. oil treated controls, # = significantly different from 3-month-old E treated group.

used in this study was such that both the A and B forms of progesterone receptor mRNA were measured concurrently.[41] Estrogen treatment increased PR mRNA levels in the MPOA of 3-month and 8-month-old rats, but was significantly less effective in inducing PR mRNA in the 15-month-old female. This corresponds to the loss of estrogen responsiveness seen at this age in terms of gonadotropin positive feedback, and to the loss of estrogen receptor binding seen with age in this same area. In contrast, there was no age-related deficit in estrogen-induced PR mRNA levels in the arcuate nucleus. Thus, these data demonstrate the ability of *in situ* hybridization histochemistry to allow semi-quantitation of mRNA levels in specific brain nuclei. The use of techniques which do not permit such fine anatomical resolution could overlook such region specific changes.

C. Ribonuclease Protection Analysis

Ribonuclease protection analysis (RPA) takes advantage of the concept that single-stranded cRNA probes will be protected from degradation by RNase if hybridized to the mRNA to which they are complementary. Consequently, by adding excess amounts of radiolabeled cRNA probe to total RNA and allowing hybridization to occur in solution, the amount of intact probe remaining following RNAse degradation will accurately represent the molar amount of specific mRNA present in the sample RNA. The intact probe is recovered from the digested solution by polyacrylamide gel electrophoresis. Furthermore, the addition of a standard curve of known amounts of *in vitro* transcribed sense RNA will allow the absolute quantitation of mRNA levels in the sample tissue. Several protocols have been published,[42] and a kit form of this assay is available (Ambion, Austin, TX).

There are several advantages and disadvantages to the use of RPA. The highly quantitative nature of RPA makes it useful for determining relatively small changes in mRNA levels. Because only a small part of the mRNA of interest is necessary to protect the radiolabeled probe, partially degraded samples may still yield good results. Ribonuclease protection assay is relatively sensitive and generally requires much smaller amounts of starting total RNA than Northern blot analysis. The fact that RPA entails solution hybridization allows for multiple probes to be added to a single RNA sample to measure several different mRNAs of interest simultaneously. Of course, if using this latter approach, the careful design of cRNA probes is essential to prevent overlap of electrophoresed bands. Finally, RPA can be useful for the determination of mRNA structure. The design of cRNA probes which overlap splice sites, areas of secondary structure or differing start sites can be used to generate multiple protected bands which reflect the amounts of each of these mRNA variants.

Ribonuclease protection analysis, however, can be difficult to control and difficult to validate. The procedure is relatively rapid, but tedious. Careful titration of the amount of RNAse used for degradation is necessary and in our hands, can vary between each mRNA of interest. Relatively clean cRNAs of a single length must be transcribed in order to prevent multiple size bands from appearing. Finally, the anatomical resolution of RPA is only as good as the microdissection procedure employed.[43]

Using an RNAse protection assay, O'Keefe et al.[33] demonstrated that transient increases (approximately 60%) in estrogen receptor-alpha mRNA occurred in the developing hippocampus. By coupling these data with *in situ* hybridization histochemistry, this study also demonstrated that these transient increases were primarily in the CA3 and CA1 pyramidal cell layer and not the dentate gyrus. Thus, this study demonstrates increases in estrogen receptor alpha mRNA levels during a time when estrogen-sensitivity is reported to be necessary for hippocampal development.

VI. Do Levels of Receptor mRNA Reflect Levels of Receptor Protein?

An assumption which is sometimes made when measuring mRNA levels is that the amount of receptor mRNA correlates with the level of receptor protein detected. Unfortunately, the relationship between mRNA levels and protein levels is complicated and depends on a variety of factors. It should be realized that mRNA levels are not static but that steady state levels of mRNA result from the dynamic processes such as mRNA synthesis and degradation. Furthermore, protein levels are a result of translational efficacy of the mRNA coupled with the rate of degradation or secretion.

A. Translational Efficacy

Initiation of translation is a complex process. *In vivo* studies using brain tissue have indirectly examined translational efficiency by the simultaneous or parallel detection of mRNA and protein levels. Non-isotopic *in situ* hybridization can be combined with isotopic autoradiography to detect both receptor mRNA and binding within the same cells.[44] Less anatomically resolute techniques can be combined as well, such as Northern blot analysis and RNAse protection analysis with receptor binding.[45,46] In general, there is a correspondence between steroid hormone receptor mRNA levels and protein or binding levels. However, some studies show that receptor mRNA levels may not correspond to protein levels.[46] Tissue specific factors may play a role in regulating translational efficacy (for review see Reference 47). Recent studies have shown that the 5'- and 3'-untranslated regions (UTRs) of mRNA are involved in regulating translation, perhaps by binding RNA binding proteins.[48] The importance of these UTRs in regulating steroid hormone receptor translation is emphasized by two findings. Steroid hormone receptor mRNAs have characteristically long UTRs and most steroid hormone receptor mRNAs seem to have multiple 5'-UTRs.

B. Multiple Forms of Receptor mRNA and Protein

All steroid hormone receptors have been shown to exhibit multiple forms of protein or mRNA. For mRNA, these differences often exist in the 5'- and 3'-UTR. Androgen receptor mRNA exists in two isoforms. The larger form is approximately 10 to 11 Kb and has been described for most tissues such as prostate gland. A second smaller form of androgen receptor mRNA exists in brain tissue (approximately 9 to 9.5 Kb)[32,34] as a result of a truncated 5'-UTR.[35] Since both mRNAs appear to have the same coding sequence, it is likely that they produce a single AR protein, yet ablation of the 5'-UTR of the larger androgen receptor mRNA may result in inefficient translation of the receptor protein, suggesting that brain tissue has adopted other ways in which to regulate the translational efficacy of the AR mRNA.

Multiple 5'-UTRs have also been identified for the mineralocorticoid receptor.[49] These alternate 5'-UTRs arise from different exons which are spliced onto a common coding region. Thus, for the mineralocorticoid receptor, the protein encoded by the different mRNAs appears to be identical. However, the efficiency with which translation occurs is dependent on the sequence of the 5'-UTR.

The estrogen receptor alpha mRNA has also been shown to exist in multiple forms as a result of at least 3 different 5'-UTRs which are spliced onto a common exon containing the translation initiation codon.[50] All of these different mRNA isoforms exist in brain, but with region specific distributions. Whether the different 5'-UTRs are responsible for brain region specific translational efficacies of the estrogen receptor remains to be determined.

In addition to multiple 5'-UTRs, the estrogen receptor is encoded by two different genes; termed α and β. These two isoforms are different in the ligand binding and transactivational domains; however, sequence homology in the DNA binding domain is high. The cloning of the β form of estrogen receptor[27] and the determination that different patterns of expression exist in brain for the α and β isoform[51] suggest the possibility that alternate forms of receptor could be found for the other steroid hormone receptors as well.

Glucocorticoid and progesterone receptor are similar in that two different isoforms of receptor have been demonstrated. For the glucocorticoid receptor, two mRNA species arise due to alternate splicing of a common gene. The protein products, termed glucocorticoid receptor α and β differ in the carboxy terminus. Heterodimerization of the GR isoforms is postulated to decrease glucocorticoid mediated transcription.[52] Similarly, progesterone receptor exists in an A and B form. The smaller A form arises from either the use of a downstream translation initiation codon,[53] or from a second promoter which initiates transcription at a later transcription initiation sequence.[54] This smaller form is less effective in initiating transcription of progesterone sensitive genes.

VII. Steroid Hormone Regulated Gene Expression as an Indicator of Sensitivity

A. Autoregulation of Hormone Receptor Expression

A well-established characteristic of steroid hormone receptor genes is their ability to be autologously regulated in brain and peripheral tissues. Thus, activation of steroid hormone receptors will up- or down-regulate receptor mRNA levels. The direction of this autologous regulation is not always the same but dependent on the receptor being examined, the brain region examined, and even the timeperiod following ligand removal. These interactions are often complex. For example, in the medial preoptic area of the rat brain, androgen receptor mRNA increases following castration, but decreases if castration is extended to several months. At both timepoints, androgen treatment will reverse the effects of castration.[55] This characteristic of steroid hormone receptor genes allows one to exploit the autoregulatory process as an index to assess the sensitivity of a given tissue to a hormone.

We have used the autoregulation of glucocorticoid receptors to examine glucocorticoid sensitivity in female rats in the presence or absence of estrogen.[19] It has been well established that in hypothalamus and hippocampus, glucocorticoid receptors are increased following adrenalectomy and decreased following glucocorticoid receptor agonist administration. We have hypothesized that estrogen treatment causes a decrease in sensitivity to glucocorticoid receptor mediated negative feedback which results in enhanced and prolonged secretion of ACTH in response to a stressor. Estrogen does not appear to alter hypothalamic or hippocampal glucocorticoid receptor binding under basal conditions. However, estrogen treatment does impair the ability of RU28362, a specific glucocorticoid receptor agonist from downregulating glucocorticoid receptor binding.[19] Thus, these results show that estrogen may impair feedback sensitivity to glucocorticoid hormones and the use of autoregulation may be one strategy by which to assess the steroid sensitivity of neuroendocrine genes.

B. Steroid Hormone Sensitive Neuroendocrine Genes

The use of autoregulation is but one strategy by which to assess the sensitivity of a given tissue. In reality, almost any steroid-sensitive target gene will suffice. Thus, this approach utilizes a given target gene as an *in vivo* reporter for sensitivity. The target gene(s) to examine should be carefully chosen. Preferably, the gene should be well characterized, and strongly and preferentially responsive to a single steroid. The function of the target gene in neuroendocrine function should be demonstrated.

For example, the use of steroid sensitive target genes as reporters for steroid sensitivity can be demonstrated by the use of the progesterone receptor as a reporter for estrogen sensitivity. Progesterone receptor is rapidly and robustly increased in hypothalamic nuclei by estrogen treatment and the amplitude of progesterone receptor induction by estrogen is reduced in some brain nuclei[56] which corresponds to the loss of behavioral or neuroendocrine effects of estrogen.

C. Factors Regulating Efficacy of Steroid Regulated Gene Expression

It is now apparent that steroid hormone receptors do not act alone in enhancing or repressing transcription of steroid hormone dependent genes. Studies examining glucocorticoid and mineralocorticoid receptor regulation of gene expression have modified

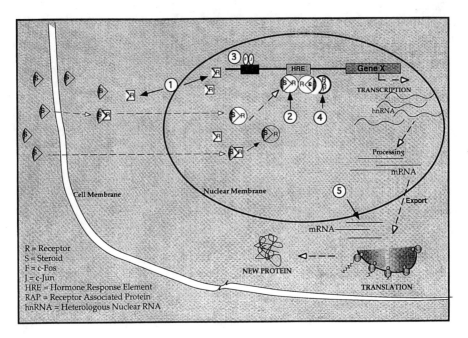

FIGURE 3

Schematic diagram of a neuroendocrine cell showing possible sites for measurement of receptor function (1,2,5) or potential sites for modifying receptor action (3,4). All are implicated in assessing the overall sensitivity of target cells to steroid hormones; 1 = number of unoccupied binding sites, 2 = number of occupied binding sites, 3 = modification by other transcription factors, 4 = modification by receptor associated proteins, 5 = receptor mRNA levels.

our thinking regarding the way in which steroid hormone receptors can interact with the transcriptional machinery. The concept of "complex response elements" in which multiple transcription factors interact to regulate receptor specificity, transcriptional gain, or whether a single steroid hormone receptor acts to enhance or repress transcription have raised interesting questions regarding the ultimate relationship of steroid hormone receptors with normal cellular functioning. Finally, several recent studies have demonstrated the presence of receptor associated proteins which can amplify or repress the actions of steroid hormone receptors[20,21] (see Figure 3).

Although these studies have utilized non-neural tissues to demonstrate the intricacies of steroid hormone dependent gene regulation, the probability is high that similar mechanisms are at work in neuroendocrine tissues. An examination of steroid receptor associated proteins, distributions in brain tissues, especially in relationship to the cells containing steroid hormone receptors, and the establishment that these proteins can interact with steroid hormone receptors to regulate neuroendocrine function may ultimately allow for the reformulation of existing hypotheses regarding regulation of the sensitivity of neuroendocrine tissues to steroid hormones.

References

1. Kawata, M., Role of steroid hormones and their receptors in structural organization of the nervous system. *Neurosci. Res.,* 24, 1–46, 1995.
2. Beatty, W. W., Hormonal organization of sex differences in play fighting and spatial behavior. *Prog. Brain Res.,* 61, 315–320, 1984.
3. Tsai, M. J. and O'Malley, B. W., Molecular mechanisms of action of steroid/thyroid receptor superfamily members. *Ann. Rev. Biochem.,* 63, 451–486, 1994.
4. Pratt, W. B., Control of steroid receptor function and cytoplasmic-nuclear transport by heat shock proteins. *BioEssays,* 14, 841–848, 1992.
5. Truss, M. and Beato, M., Steroid hormone receptors: interaction with deoxyribonucleic acid and transcription factors. *Endocr. Rev.,* 14, 459–479, 1993.
6. Funder, J. W., Pearce, P. T., Smith, R., and Smith, A. I., Mineralocorticoid action: target tissue specificity is enzyme, not receptor mediated. *Science,* 242, 583–585, 1988
7. Pearce, D., A mechanistic basis for distinct mineralocorticoid and glucocorticoid receptor transcriptional specificities. *Steroids,* 59, 153–159, 1994.
8. Dallman, M. F. and Yates, F. E., Dynamic assymmetries in the corticosteroid feedback pathway and distribution binding and metabolism of elements of the adrenocortical system. *Ann. N.Y. Acad. Sci.,* 156, 696–721. 1969.
9. Widmaier, E. P. and Dallman, M. F., The effects of corticotropin-releasing factor on adrenocorticotropin secretion from perifused pituitaries *in vitro*: rapid inhibition by glucocorticoids, *Endocrinology,* 115, 2368–2374, 1984.
10. Saphier, D. and Feldman, S., Iontophoretic application of glucocorticoids inhibits identified neurones in the rat paraventricular nucleus, *Brain Res.,* 453, 183–190, 1988.
11. Teyler, T., Vardis, R. M., Lewis, D., and Rawitch, A. B., Gonadal steroids: effects on excitability of hippocampal pyramidal cells. *Science,* 209, 1017–1019, 1980.
12. Wong, M. and Moss, R. L., Long-term and short-term electrophysiological effects of estrogen on the synaptic properties of hipocampal CA1 neurons. *J. Neurosci.,* 12, 3217–3225, 1992.
13. Belelli, D., Lan, N. C., and Gee, K. W., Anticonvulsant steroids and the GABA/benzodiazepine receptor-chloride ionophore complex. *Neurosci. Biobehav. Rev.,* 14, 315–322, 1990.
14. Towle, A. C. and Sze, P. Y., Steroid binding to synaptic plasma membrane: differential binding of glucocorticoids and gonadal steroids. *J. Steroid Biochem.,* 18, 135–143, 1983.
15. Orchinik, M., Murray, T. F., and Moore, F. L., A corticosteroid receptor in neuronal membranes. *Science,* 252, 1848–1851, 1991.
16. Ramirez, V. D., Zheng, J., and Siddique, K. M., Membrane receptors for estrogen, protesterone and testosterone in the rat brain: fantasy or reality. *Cell. Mol. Neurobiol.,* 16, 175–198. 1996.
17. Gorski, R. A., Nature of hormone action in the brain. In: *Ontogeny of Receptors and Reproductive Hormone Action.* Hamilton, T. H., Clark, J. H., and Sader, W. A. (Eds.), Raven Press, New York, 371–392. 1979.

18. Brown, T. J., Hochberg, R. B., Zielinski, J. E., and MacLusky, N. J., Regional sex differences in cell nuclear estrogen binding capacity in the rat hypothalamus and preoptic area. *Endocrinology,* 123, 1761–1770, 1988.

19. Burgess, L. H. and Handa, R. J., Estrogen alters adrenocorticotropic hormone and corticosterone secretion and glucocorticoid receptor mediated function. *Endocrinology,* 131, 1261–1269, 1992.

20. Yeh, S. and Chang, C., Cloning and characterization of a specific coactivator, ARA70, for the androgen receptor in human prostate cells. *Proc. Natl. Acad. Sci. USA,* 93, 5517–5521, 1996.

21. Smith, C. L., Onate, S. A., Tsai, M-J., and O'Malley, B. W., CREB binding protein acts synergistically with steroid receptor coactivator-1 to enhance steroid receptor dependent transcription. *Proc. Natl. Acad. Sci. USA,* 93, 8884–8888, 1996

22. Handa, R. J., Reid, D. L., and Resko, J. A., Androgen receptors in brain and pituitary of female rats: cyclic changes and comparisons with the male. *Biol. Reprod.,* 34, 293–304. 1986.

23. Reul, J. M. H. M., Van Den Bosch, F. R., and De Kloet, E. R., Relative occupation of type-I and type-II corticosteroid receptors in rat brain following stress and dexamethasone treatment: functional implications. *J. Endocrinol.,* 115, 459–467, 1987.

24. Brown, T. J., Sharma, M., Heisler, L. E., Karsan, N., Walters, M. J., and MacLusky, N. J., *In vitro* labeling of gonadal steroid hormone receptors in brain tissue sections. *Steroids,* 60, 726–737, 1995.

25. Paul, S. J., Ortolano, G. A., Haisenleder, D. J., Stewart, J. M., Shupnik, M. A., and Marshall, J. C., Gonadotropin subunit messenger RNA concentration after blockade of gonadotropin releasing hormone action: Testosterone selectively increases follicle stimulating hormone beta-subunit messenger RNA by postranscriptional mechanisms. *Mol. Endocrinol.,* 4, 1943–1955. 1990.

26. Wilson, M. E. and Handa, R. J. Regulation of follicle stimulating hormone mRNA in primary cultures of infantile female rat pituitary cells by activin, androgen and GnRH. *Abstr. Endocrine Society Ann. Mtg.,* 212, 1996.

27. Kuiper, G. G., Enmark, E., Pelto-Huikko, M., Nilsson, S., and Gustafsson, J. A., Cloning of a novel estrogen receptor expressed in rat prostate and ovary. *Proc. Natl. Acad. Sci. USA,* 93, 5925–2930, 1996.

28. Stumpf, W. E., Sar, M., and Keefer, D. A., Atlas of estrogen target cells in rat brain. In *Anatomical Neuroendocrinology,* Karger, Basel, 104–119, 1975.

29. Chang, C., Kokontis, C., and Liao, S., Molecular cloning of human and rat complementary DNA encoding androgen receptors. *Science,* 240, 324–326, 1988.

30. Green, S., Walter, P., Kumar, V., Krust, A., Bornert, J. M., Argos, P., and Chambon, P., Human oestrogen receptor cDNA: sequence, expression and homology to v-erb-A. *Nature,* 320, 134–139, 1986.

31. Arriza, J. L., Weinberger, C., Cerelli, G., Glaser, T. M., Handelin, B. L., Housman, D. E., and Evans, R. M., Cloning of human mineralocorticoid receptor complementary DNA: Structural and functional kinship with the glucocorticoid receptor. *Science,* 237, 268–275, 1987.

32. Burgess, L. H. and Handa, R. J., Hormone regulation of androgen receptor messenger ribonuclieic acid in brain and pituitary of male rat. *Mol. Brain Res.,* 19, 31–38, 1993.

33. O'Keefe, J. A., Li, Y., Burgess, L. H., and Handa, R. J., Estrogen receptor mRNA alterations in the developing rat hippocampus. *Mol. Brain Res.*, 30, 115–124, 1995.

34. McLachlan, R., I., Tempel, B. L., Miller, M. A., Bicknell, J. N., Bremner, W. J., and Dorsa, D. M., Androgen receptor gene expression in the rat CNS: evidence for two mRNA transcripts. *Mol. Cell. Neurosci.*, 2, 117–122. 1991.

35. Price Jr., R. H., Wilson, M. E., and Handa, R. J., A smaller form of androgen receptor mRNA in brain contains a truncated 5'-untranslated region. *Soc. Neurosci. Abstr.*, 21, 882, 1995

36. Mizokami, A. and Chang, C., Induction of translation by 5'-untranslated region of human androgen receptor mRNA. *J. Biol. Chem.*, 269, 25655–25659, 1994.

37. Simerly, R. B., Chang, C., Muramatsu, M., and Swanson, L. W., Distribution of androgen and estrogen receptor mRNA-containing cells in the rat brain: an *in situ* hybridization study. *J. Comp. Neurol.*, 294, 76–95, 1990.

38. Simmons, D. M., Arriza, J. L., and Swanson, L. W., A complete protocol for *in situ* hybridization of messenger RNAs in brain and other tissues with radiolabeled single-stranded RNA probes. *J. Histotechnol.*, 12, 169–181, 1989.

39. Hrabovsky, E., Vrontakis, M. E., and Petersen, S. L., Triple labelling method combining immunocytochemistry and *in situ* hybridization histochemistry: demonstration of overlap between Fos-immunoreactive and galanin mRNA expressing subpopulations of luteinizing hormone releasing hormone neurons in female rats. *J. Histo. Cytochem.*, 43, 363–370, 1995.

40. Scarbrough, K., Weiland, N. G., Larson, G. H., Sortino, M. A., Chiu, S. F., Hirschfield, A. N., and Wise, P. M., Measurement of peptide secretion and gene expression in the same cell. *Mol. Endocrinol.*, 5, 134–142, 1991.

41. Kato, J., Hirata, S., Nozawa, A., and Mouri, N., The ontogeny of gene expression of progestin receptors in the female rat brain. *Mol. Biol.*, 47, 173–182, 1993.

42. Kinloch, R. A., Roller, R. J., and Wasserman, P. M., Quantitative analysis of specific messenger RNAs by ribonuclease protection. *Meth. Enzymol.*, 225, 294–303, 1993.

43. Kerr, J. E., Allore, R. J., Beck, S.G., and Handa, R. J., Distribution and hormonal regulation of androgen receptor messenger ribonucleic acid in rat hippocampus. *Endocrinology*, 136, 3213–3221, 1995

44. Toran-Allerand, C. D., Miranda, R. C., Hochberg, R. B., and MacLusky, N. J., Cellular variations in estrogen receptor mRNA translation in the developing brain: evidence from combined [^{125}I]estrogen autoradiography and non-isotopic *in situ* hybridization histochemistry. *Brain Res.*, 576, 25–41, 1992.

45. Bohn, M. C., O'Banion, M. K., Young, D. A., Giuliano, R., Hussain, S., Dean, D. O., and Cunningham, L. A., *In vitro* studies of glucocorticoid effects on neurons and astrocytes. *Ann. N.Y. Acad Sci.*, 746, 243–258, 1994.

46. Yang, K., Hammond, G. L., and Challis, J. R., Characterization of an ovine glucocorticoid receptor cDNA and developmental changes in its mRNA levels in the fetal sheep hypothalamus, pituitary gland and adrenal. *J Mol. Endo.*, 8, 173–180, 1992.

47. Jackson, R. J., Cytoplasmic regulation of mRNA function: the importance of the 3' untranslated region. *Cell*, 74, 9–14, 1993.

48. Gray, N. K. and Hentze, M. W., Regulation of protein synthesis by mRNA structure. *Molec. Biol. Rep.*, 19, 195–200, 1994.

49. Kwak, S. P., Patel, P. D., Thompson, R. C., Akil, H., and Watson, S. J., 5'-Heterogeneity of the mineralocorticoid receptor messenger ribonucleic acid: differential expression and regulation of splice variants within the rat hippocampus. *Endocrinology,* 133, 2344–2350, 1993.

50. Hirata, S., Koh, T., Yamada-Mouri, N., Hoshi, K., and Kato, J., The untranslated first exon "exon OS" of the rat estrogen receptor (ER) gene. *FEBS Lett.,* 394, 371–373, 1996.

51. Shugrue, P. J., Komm, B., and Merchenthaler, I., The distribution of estrogen receptor-β mRNA in the rat hypothalamus. *Steroids,* 61, 678–681, 1996.

52. Bamberger, C. M., Bamberger, A. M., DeCastro, M., and Chrousos, G. P., Glucocorticoid receptor beta, a potential endogenous inhibitor of glucocorticoid action in humans. *J. Clin. Invest.,* 95, 2435–2441, 1995.

53. Conneely, O. M., Maxwell, B. L., Toft, D. O., Schrader, W. T., and O'Malley, B. W. The A and B forms of the chicken progesterone receptor arise by alternate initiation of translation of a unique mRNA. *Biochem. Biophys. Res. Commun.,* 149, 493–501, 1997.

54. Kastner, P., Krust, A., Turcotte, B., Stropp, U., Tora, L., Gronemeyer, H., and Chambon, P., Two distinct estrogen-regulated promoters generate transcripts encoding the two functionally different human progesterone receptor forms A and B. *EMBO J.,* 9, 1603–1614, 1990.

55. Handa, R. J., Kerr, J. E., DonCarlos, L. L., McGivern, R. F., and Hejna, G., Hormonal regulation of androgen receptor messenger RNA in the medial preoptic area of the male rat. *Mol. Brain Res.,* 39, 57–67, 1996.

56. Brown, T. J., MacLusky, N. J., Shanabrough, M., and Naftolin, F., Comparison of age- and sex-related changes in cell nuclear estrogen-binding capacity and progestin receptor induction in the rat brain. *Endocrinology,* 126, 2965–2972, 1990.

Chapter

In Vitro Approaches to Neuroendocrinology

Celia D. Sladek

Contents

0-8493-3363-6/98/$0.00+$.50
© 1998 by CRC Press LLC

A variety of *in vitro* approaches have provided diverse insights into the function of neuroendocrine systems. These range from long and short term explant preparations that maintain tissue morphology to long and short term dispersed cell preparations. The explant cultures have been particularly useful for studies of hormone release due to maintenance of axonal projections to the median eminence and posterior pituitary. The dispersed prepartions have proved useful for studies addressing developmental issues and questions related to regulation of neuroendocrine gene expression. Specific examples of these types of preparations and their distinguishing characteristics will be presented in order to provide insights into the types of questions that can be addressed in each preparation.

I. Explant Preparations

A. Short Term

Explant preparations consisting of microdissected portions of the hypothalamus have been used to study release of a variety of neurohormones including vasopressin (VP), oxytocin (OT), luteinizing hormone releasing hormone (LHRH), corticotropin releasing hormone (CRH), cholecystokinin (CCK), β-endorphin, neurotensin, and dopamine (DA)[1-12] as well as circadian rythmicity in the suprachiasmatic nucleus (SCN). These explants can be maintained for hours to days in either static or perifusion conditions. A major advantage of these preparations in addition to the maintenance of three dimensional tissue morphology mentioned above is the ability to harvest tissue from adult animals. This guarantees that responses observed reflect those of a fully mature neuroendocrine system and therefore can be anticipated to reflect functions observed *in vivo* in adult animals.

1. Explants of the Hypothalamo-Neurohypophyseal System (HNS)

Several laboratories have utilized HNS explants to study VP and OT release,[1,13-29] VP and NP synthesis,[2,30] VP and OT gene expression,[31-34] and electrophysiological characterization of VP and OT neurons.[35-37] Although some of the originial studies used larger portions of the hypothalamus,[13,14] the majority of these studies utilized explant preparations consisting of a horizontal slice from the ventral hypothalamus approximately 2mm thick that includes the bilateral supraoptic nucleus (SON), the median eminence, pituitary stalk, and the attached neural lobe. Thus, the explant maintains intact the magnocellular neurons in SON that project to posterior pituitary and release VP and OT into the blood for transport to distant target tissue. Also included in the explant are the arcuate, suprachiasmatic, and portions of the anterior and ventromedial hypothalamic nuclei as well as the perinuclear zone immediately adjacent to SON and the organum vasculosum of the lamina terminalis (OVLT) and surrounding tissue. However, in order to minimize explant thickness, the paraventricular nucleus is excluded. Several of these areas innervate the HNS and provide afferents important for regulation of VP and/or OT release. Specifically, the perinuclear zone contains neurons with a variety of transmitter phenotypes that innervate SON. This zone is thought to function as an integrating center for a variety of afferent information because it is innervated by afferents from regions known to transduce information about fluid, electrolyte, and cardiovascular homeostatsis.[38] The OVLT region contains osmoreceptive elements required for osmotic regulation both *in vivo* and in HNS explants.[36,39,40] DA and β-endorphin neurons in the arcuate nucleus innervate the neural lobe and SON respectively.[41,42] Thus, in addition to maintaining the integrity of the VP and OT neurons, the HNS explant maintains several hypothalamic afferent connections to these neurons. This has contributed to the utility of this preparation for studies on the regulation of VP and OT release, and the importance of the integrity of these connections for certain responses has been confirmed by utilizing explants from animals prepared with lesions of selected regions (e.g., OVLT[39]).

a.   **Studies of hormone release from HNS explants.**   Hormone release from HNS explants has been studied utilizing static incubation, static incubation in specialized chambers that separate the medium incubating hypothalamus from that incubating neural lobe,[19] and in perifusion culture. In all of the approaches, the explants continue to release hormones into the culture medium and maintain responsiveness to physiological stimuli for several days *in vitro*. This reflects survival of sufficient SON neurons in the explant to evoke detectable responses. The survival of these neurons for long periods in the explant is probably promoted by the location of the SON on the ventral surface of the hypothalamus thus allowing maximal exposure to oxygen and nutrients provided in the culture medium. Also, since the SON neurons are on the uncut surface of the explant, the neurons are exposed to relatively little trauma during explant preparation. The static incubation approach has provided insights into the regulation of VP release by acetylcholine, norepinephrine, angiotensin II, GABA, prostaglandins, and osmotic stimulation[1,15-17,19,20] as well as the role of intracellular signaling mechanisms in mediating these responses.[22]

The use of divided chambers for static incubation studies allowed investigators to determine whether the site of action of a given agent (e.g., neurotransmitter, neuropeptide, osmotic stimulus) was in the hypothalamus or directly on the nerve terminals in posterior pituitary. This approach is particularly appropriate for evaluating the site of action of agents that innervate both the SON and the posterior pituitary (e.g., DA, GABA, NE, glutamate) or agents that may have selective access to these sites as a result of their production in the periphery and the lack of a blood brain barrier (BBB) in the posterior pituitary. Thus, using the "compartmentalized" HNS explant, Gregg[43] demonstrated that VP release from the neural lobe could be stimulated by activation of nicotinic receptors in the hypothalamus and muscarinic receptors in posterior pituitary, but that intact axonal connections between hypothalamus and pituitary were required for acetylcholine to elicit VP release via a direct effect on the neural lobe. Rossi[28] demonstrated that the site of action of endothelin-3 stimulation of VP release is in the neural lobe, not the hypothalamus. This is significant, because endothelin-3 generated in the periphery could access the VP nerve terminals in posterior pituitary without crossing the BBB. The compartmentalized approach was also used successfully to demonstrate a hypothalamic site of kappa opioid inhibition of VP release.[44]

Although there are still many interesting questions to address with the divided chamber protocol, the tedious nature of the experiments has limited its use. Instead investigators have studied hormone release from independent explants of the posterior pituitary or hypothalamus. As demonstrated by Gregg's finding,[43] studies on hormone release from the isolated posterior pituitary is limited by the viability of the preparation due to separation of the axon terminals from the cell bodies, but acute studies evaluating hormone release in response to electrical stimulation of the cut stalk have provided important insights into the ability of agents to modify stimulus-secretion coupling at the nerve terminal[45-49] as well as the importance of patterned action potential generation for hormone release.[50-52] Hypothalamic explants without the attached neural lobe have also been used in studies on VP release.[7,10,53-55] In the majority of these studies, the hypothalamic explant has included the paraventricular nucleus (PVN), and the goal has been to evaluate peptide release from the PVN-median eminence axis (see Section I.A.4.). This approach has been useful, but the multiple sites of VP release in these explants make observed effects on VP release difficult to interpret. In the HNS explant, the amount of VP released from the posterior pituitary far exceeds that released from hypothalamus.[19] As a result, experimentally induced changes in VP release can be attributed to release from the neural lobe, but in the hypothalamic explants the amount of VP released is smaller and could result from release from terminals in median eminence, dendrites in SON and PVN,[19,56-60] neurons in SCN (see below), as well as cut axons of cells projecting to neural lobe. Thus, the relative contribution of each of these components to observed responses is an issue.

The perifusion approach for culturing HNS explants is more amenable to the evaluation of the time course of responses and to the evaluation of the effect of sequential application of agents than the static incubation method. Also, it provides the unique capability to manipulate the rate of delivery of a stimulus. This has been extremely useful in studying osmotic regulation of VP and OT release.[23,33]

b.   Studies on regulation of VP mRNA in HNS explants.   Although the content of VP mRNA decreases substantially in cultured HNS explants,[31,32] it has been possible to demonstrate regulation of VP and OT mRNA content in these explants by comparision to mRNA content of explants maintained simultaneously under control conditions.[32,34,61] VP and OT mRNA content is increased in response to a slow increase in osmolality extended over 6 hr or longer, but not in response to a step increase in osmolality.[32] This has allowed evaluation of the role of synaptic activation, excitatory amino acid (EAA) neurotransmission, and steroid hormones in the osmotic regulation of VP mRNA.[29,32,34] The role of cAMP in the regulation of VP mRNA content has also been studied.[61]

c. Electrophysiological studies on HNS explants. HNS explants have been useful for extracellular and intracellular recordings for electrophysiological studies on SON neurons. In these studies the explants were either maintained in an open perifusion system[35] or were perfused via a glass cannula placed in the anterior cerebral artery.[62] The open nature of the preparations allowed placement of stimulation and recording electrodes in various regions of the explant. That aspect combined with maintenance of the hypothalamo-neurohypophyseal pathway allows neurohypophyseal neurons to be identified by antidromic activation following electrical stimulation of the neural stalk prior to the recording session. The placement of an additional stimulating electrode in the anterior hypothalamus allows for evaluation of responses to synaptic activation from the preoptic region.[62] Neurons in these preparations continue to display electrical activity for at least 15 hr. Thus, the preparation provides adequate time for extensive experimentation. Studies with these preparations elucidated several characteristics of electrical activity in SON neurons;[63-67] allowed comparisons between electrical activity and hormone release;[68] clearly demonstrated the role of alpha-adrenergic receptors in inducing action potentials in SON neurons;[63,69] originally identified the ability of histamine to potentiate electical activity induced by other excitatory mechanisms;[70] confirmed the inherent osmosensitivity of SON neurons and the importance of excitatory input from the preoptic region for osmotic activation of SON neurons;[65,71] and evaluated the role of inhibitory and excitatory afferents on SON neurons.[37,67] Thus, the combination of electrophysiological and HNS explant techniques has provided extensive insights into the function of this neuroendocrine system.

2. SCN Explants for Circadian Clock Studies

a. Neuropeptide release. Both static and perifusion approaches have also been used to examine circadian rhythmicity in neuropeptide release from explants of the SCN.[72,73] In this case, the perifusion approach is beneficial due to the extended time frame required for analysis of circadian rhythms. The explants used in these studies were microdissected from adult hypothalamus and included only the bilaterally paired SCN, their rostral projections to the OVLT, and the underlying optic chiasma (see Figure 1 in Reference 72). Individual explants release VP into the culture medium in a circadian fashion for at least 4 cycles. Regardless of whether explants are prepared near the onset or end of the light phase of the light:dark cycle, VP release was characterized by peak levels during the subjective day and dark levels during the subjective night. Thus, the explants retain the rhythmic pattern characteristically observed *in vivo*. This observation confirmed the presence of one or more circadian oscillator(s) within the explant, and has provided the opportunity to study the physiology of this circadian pacemaker.[73,74] Although many investigators have turned to hypothalamic slice cultures for *in vitro*, organotypic studies on SCN (see Section I.B.2.), due to their longevity *in vitro*, the explants described here retain the advantage that they are obtained from young adult animals and therefore, will reflect the characteristics of a fully mature circadian system. This may not be true for the SCN in hypothalamic slice cultures derived from prenatal animals.[75]

b. Electrophysiological studies on SCN neurons in vitro. *In vitro* approaches have also been used to evaluate the electrophysiological characteristics of the SCN.[76-78] In this case, the SCN was included in a coronal slice of hypothalamus from the level of the optic chiasma. Neurons could be recorded from these preparations for up to 53 hr[79] and therefore could be studied over several ciradian cycles. The results confirmed the maintenance of circadian periodicity in electrical activity in SCN neurons that corresponded with the fluctuations in VP release described above. This preparation was subsequently used to demonstrate a role for cAMP and serotonin in resetting the clock[79,80] as well as effects of GABA and anxiolytics on single unit discharge of SCN neurons.[78]

3. Preoptic/Median Eminence Explants for Studies of LHRH Release

The success of studies using the HNS explants led investigators to devise hypothalamic explant preparations appropriate for studying the other hypothalamic neuroendocrine systems. The primary consideration in preparing these explants was to dissect a small portion of the hypothalamus that included the neuronal perikarya, axon, and terminal. For the neuroendocrine systems regulating anterior pituitary function, the terminal field was the median eminence. Therefore, the intent was to maintain the integrity of the axonal projection from the perikarya in a small enough piece of tissue that the neurons of interest would receive adequate nutrient and oxygen delivery to sustain viability. For studies on LHRH release using rat tissue, this meant dissecting a medial piece of hypothalamus that extended rostrally to include the preoptic area (POA) and caudally to the end of the median eminence.[4,81] Longitudinal bisection of the tissue block allowed exposure of the uncut surface of the third ventricle for diffusion of oxygen and nutrients. Immunocytochemical studies demonstrated that these explants contained the LHRH neurons in POA and their axons terminating both in the OVLT and the median eminence.[4] When maintained in perifusion culture, these explants released LHRH into the medium for at least 26 hr and LHRH release could be stimulated by the addition of KCl at the end of the perifusion.[4] Short interval collections of the perifusate demonstrated that the LHRH release occurred in a pulsatile manner.[82,83] Since these explants could be prepared from donors of different ages, they were useful for studies on the peripubertal control of LHRH release[4,81,82] as well as evaluation of the interaction of various neuroactive agents in regulation of LHRH release.[84-86]

4. Explants for Studies of CRH/VP Release

As described above, hypothalamic explants designed for evaluating hypothalamic release of CRH and co-localized VP have been used extensively.[5,7,87] These explants differ from the HNS explants in that they do not include the neural lobe and extend deeper into the hypothalamus to include the PVN. They have been useful in identifying neurotransmitters that regulate CRH release;[7,87] in comparing effects of these agents on CRH and VP release;[88] in evaluating interaction of these agents and corticosteroids in regulation of CRH release;[89] and in demonstrating interactions between the immune system and CRH release.[88]

5. Explants for Studies of SRIF/GHRH Release

The anatomy of the somatostatin (SRIF) and growth hormone releasing hormone (GHRH) neurons make this system particularly amenable to studies using hypothalamic explant preparations, because both sets of neurons are localized close to the third ventricle and their axons project a relatively short distance to the median eminence.[90] The SRIF neurons are located in the periventricular zone and the GHRH neurons in the arcuate nucleus. This provides interesting possibilties for studies on the comparative regulation of these two neuroendocrine peptides that have opposing actions on the regulation of growth hormone release from anterior pituitary. The explants used in these studies include the medial hypothalamic tissue (0.6 mm lateral to the third ventricle) posterior to the optic chiasma and either anterior to the mammillary bodies[91] or just including anterior median eminence.[92,93] They are cut at a depth of 2 mm to allow inclusion of the SRIF neuronal perikarya or at a depth of 1 mm to exclude the PVN.[94] The thicker explants are bisected longitudinally. This allows the third ventricle to lie open and improves nutrient and oxygen diffusion to the SRIF and GHRH neurons. These explants have been used to demonstrate a hypothalamic site for the negative feedback action of growth hormone to stimulate SRIF release and induce SRIF mRNA, and in conjunction with incubations of median eminence fragments alone to demonstrate that growth hormone acts directly on the median eminence to inhibit GHRH release.[92] They have also been used to evaluate the effect of insulin-like growth factor I on SRIF release,[93,94] and to demonstrate altered regulation of SRIF, GHRH, and thyrotropin releasing factor by glucose in explants from diabetic rats.[92]

6. Explants of the Tuberohypophyseal Dopamine System

Since the tuberohypophyseal dopamine neurons that innervate the intermediate lobe for regulation of secretion of the pro-opiomelanocortin peptides as well as those that release dopamine into the portal vasculature for inhibition of prolactin release from the anterior pituitary are located in the arcuate nucleus, similar explants to those described above for studies on SRIF and GHRH have been used to study regulation of dopamine release. In this case, the explant included the mediobasal hypothalamus with the median eminence, infundibular stalk and attached neurointermediate lobe.[3,95] Differential release of dopamine and serotonin from these explants was observed in response to electrical stimulation and neurotensin.[3]

B. Long-Term Explant Cultures

In contrast to the explants described above that are useful for studying neuroendocrine systems over a period of hours to days, other types of hypothalamic explants prepared from fetal tissues have proved useful for studies that demand stable systems lasting for several weeks. Two approached will be described: Explants of selected hypothalamic regions punched from fetal, neonatal, or adult brain, and slice explants prepared from sections of fetal rat brain and maintained in either stationary or roller tube culture. These techniques represent the earliest successful approaches for extended maintenance of hypothalamic tissue *in vitro* (for review see Reference 96) and are still widely used to address a variety of issues in neuroendocrinology.

1. Punch Explants

The first reports of successful maintenance of hypothalamic neurons in culture appeared in the early fifties.[96] In these studies, neurons that grew out of small fragments of the PVN and SON maintained in plasma clots were viable for several weeks. The early studies primarily contributed morphological characterization of the cultures, but as advanced electrophysiological techniques became available and immunological approaches were developed for identifying the hypothalamic neurons in culture and for studying peptide release, these cultures proved useful for a variety of experiments on neuroendocrine systems. Sakai et al.[97] used SON explant punches from neonatal puppies to characterize the electrophysiological effects of a variety of neurotransmitters using intracellular recording techniques, and Gawhiler et al.[98] used SON explants from newborn rats to demonstrate synchronized firing of SON neurons characteristic of OT neurons during suckling as well as the phasic firing pattern characteristic of VP neurons. They also demonstrated the presence of synapses on the cultured cells by intracellular recording of postsynaptic potentials.[98] Sakai et al.[97] demonstrated VP immunoreactivity in the explant, but was not able to detect VP release into the culture medium. However, Morris and colleagues[99] utilized explants of SON and PVN prepared from either neonatal or young rats to compare VP and OT synthesis and release between normotensive and spontaneously hypertensive rats.

2. Organotypic Slice Explants

Slice explants are generally prepared as 400 µm thick slices from embryonic or neonatal donors and maintained in either Maximov depression slide chambers or on cover slips in roller tubes. The Maximov assembly has the advantages of direct observation of the explant during culture and maintaining 3 dimensional tissue organization.[100] This approach has been useful in developmental studies including the role of steroids in the sexual differentiation of the hypothalamus.[101,102] In the roller tube technique, the cultures maintain a high degree of cytological organization, but tend to flatten to near monolayer thickness. This provides the ability to visualize all parts of a given neuron for intracellular or patch clamp electrophysiological recordings, to make intracellular injections for anatomical characterization of individual neurons, or to visualize changes in intracellular calcium in individual cells.[100,103-105] The neurons can be tentatively identified by their location in the slice prior to recording, or they can be identified by their projection pathway by labelling with flourescent retrograde tracers prior to culture.[104] The slices are amenable to immunocytochemistry and *in situ* hybridization techniques.[104] Other major advantages of the roller tube technique are the ease of maintenance of the slices for extended periods of time and the ability to prepare co-cultures to assess the role of innervated brain regions on neuronal development.[103,104,106] The ability to maintain these slice cultures in serum free medium further enhances their utility, because they can be used to evaluate regulation of neuronal activity in a highly controlled environment.[107] The entire hypothalamus from 7-day-old rats can be maintained in culture by dividing it into coronal sections 400 μm thick.[103] Each section contains unique components of the hypothalamus.[106] Thus, preoptic sections are useful for studies on LHRH neurons, but more caudal sections are useful for studies on SCN or the neurohypophyseal magnocellular system, and even more caudal sections for studies on tyrosine hydroxylase (TH) neurons in the arcuate nucleus.

Hypothalamic slice cultures have been used extensively for studies on the SCN, magnocellular OT neurons, and the LHRH neurons.[75,100,103,104,106,107] They have also been shown to contain immunoreactive neurotensin, TH, and CRH neurons.[106] However, magnocellular VP neurons are virtually absent in slice cultures maintained beyond 2 weeks.[107] This is in contrast to the impressive survival of parvocellular VP neurons in the SCN,[107] and raises interesting questions relative to selective requirements for neuronal survival or gene expression in the magnocellular VP compared with OT neurons. Corticosterone inhibition of CRH has been demonstrated in short term hypothalamic cultures.[108] LHRH neurons demonstrate similar morphological differentiation in slice cultures and *in vivo*,[106] and co-cultures of preoptic hypothalamic slices with anterior pituitary demonstrated the importance of LHRH neurons for maintenance of LH expression in anterior pituitary cultures. However, preoptic slices cultured with brainstem did not establish synaptic connections between LHRH neurons and brainstem TH neurons.[106]

The SCN maintained in cultured slices contains neurons that express vasoactive intestinal polypeptide (VIP), gastrin releasing hormone (GRP), and GABA as well as VP,[75,109] however during culture, the VP and VIP neurons develop in a manner similar to that observed *in vivo*, but development of the GRP neurons in the slice cultures is attenuated.[75] The SCN maintained in slice culture retains the ability to generate circadian rhythms as demonstrated by the existence of circadian patterns in the release of VP and VIP from slice explants containing the SCN.[110,111] In sights into the function of the circadian pacemaker have been obtained in experiments with SCN slice cultures. For example, the phase of the VP release pattern from slice cultures can be shifted by treatment with protein synthesis inhibitors, and treatment with NMDA shifts the rhythms in VP and VIP release independently suggesting the existence of independent circadian oscillators.[110,112,113] Thus, hypothalamic slice cultures have proved useful for a variety of studies relevant to neuroendocrine systems.

II. Dispersed Hypothalamic Cell Preparations

A. Primary Cultures of Hypothalamic Neurons

Primary hypothalamic cultures have been used to study essentially all neurons involved in neuroendocrine function including magnocellular VP and OT neurons, the parvocellular neurons producing LHRH, CRH, SRIF, GHRH, β-endorphin, and TH, and neurons of the SCN. Most frequently, these primary cultures are prepared by enzymatic digestion and disruption of fetal hypothalamic tissue followed by plating in a variety of culture vessels using some version of modified Eagle's medium either in the presence of absence of serum. The conditions providing for optimal cultures vary with the neuron of interest and the experimental paradigm, but issues that impact on establishing optimal cultures include age of the donor, plating conditions, and the components of the culture medium. Therefore, each of these will be discussed in some detail relative to their impact on the characteristics of the resultant culture.

1. Characteristics of Primary Hypothalamic Cultures

Since hypothalamic neurons are derived from the neural stem cells lining the third ventricle, cultures of fetal hypothalamic neurons are usually established from periventricular tissue regardless of the location of the neurons of interest in the mature brain.[114] As a result, these primary cultures contain heterogeneous cell types including a wide variety of neuronal phenotypes as well as glial cells. In general the glial cells are beneficial for maintenance of the neuronal elements, because they proliferate to form a cell carpet that is ideal for neuronal attachment and extension of neuritic processes. However, conditions that promote neuronal survival (e.g., the inclusion of serum in the culture medium) also promote glial proliferation, and this can be detrimental to long term survival of neurons if the glia overgrow the neurons. As a result, many investigators employ antimitotic agents (e.g., cytosine arabinoside) to control glial proliferation.[114,115] The co-existence of multiple neuronal populations in hypothalamic cultures has been demonstrated by several investigators.[114-123] Depending on the conditions used, the culture may contain both magnocellular and parvocellular VP neurons,[114,115,117-120,124-131] as well as neurons producing OT,[132] SRIF,[114,117,118,121-123,133-135] GHRH,[121,136] CRH,[117,118,137] LHRH,[116,138,139] TH,[114,119,140] GABA,[40,122,123,141] opioid peptides and pro-opiomelanocortin,[115,120,142,143] neuropeptide Y,[115] atrial natriuretic peptide,[144] and nitric oxide synthase.[115]

2. Donor Age

a. Use of fetal or embryonic tissue. The optimal donor age varies depending upon the cell type of interest and the experimental objective, but is usually at or shortly after the time of 'birth' of the neuronal cell group of interest. During this period *in vivo*, the neurons are generated at the proliferative zone, migrate to their final location in the hypothalamus, and begin extending axonal and dendritic projections. The dispersion protocol is thought to be less traumatic during this early period of development for any given group of neurons, because fewer processes are disrupted. As shown in Table 1, embryonic day 14 to 19 (E14-E19) is most commonly used to establish long term cultures of rat hypothalamus. However, within this range there is variability based on the specific neuron of interest, and older tissue (postnatal and even adult) has been used (see below). The optimal donor age

TABLE 1
Age of Donor Used to Establish Primary Hypothalamic Cultures

Neuronal phenotype	Presence (+) or absence (−) of substance	Donor age
VP	+	E13-15[125]
	+	E14[114,126,127,129]
	+	E14-15[115]
	+	E15[126,128]
	+	E17[126,128,131]
	+	E18[116,117,124,126,128]
	+	E19[119,124,128,131]
	+	E19-22[130]
	+	Adult[148]
OT	−	E14[114]
	−	E15[128]
	−	E16[120,128,132]
	−	E17[132]
	+[a]	E18[132]
	+[a]	E19[132]
	+	P0-P2[132]
SRIF	+ (few cells)	E14[114]
	+	E17[121,133]
	+	E18[117,118,122]
	+	Adult[149]
GHRH	+	E17[121,136]
CRH	+	E18-19[117,124]
	+	E21[137]
LHRH	+	E17[139]
	+	E18-19[116,124]
	+	P6[138]
	+	Adult[150]
TH	+	E14[114]
	+	E19-20[140]
TRH	+	E17[133]
GABA	+	E14-15[115,141]
	+	E17[123]
β-endorphin	+	E17[142]
	+	P2[143]

Note: E, embryonic; P, postnatal.

[a] When treated with triiodothyronine.

has not been systematically investigated for all types of neuroendocrine neurons, but it was evaluated for VP neurons in primary cultures by DiScala et al.[128] and Schilling and Pilgrim.[126] DiScala found that although VP neurons were present in cultures from established from fetuses ranging in age from E15 to E19, the period of survival of magnocellular neurons was attenuated in cultures from the older fetuses (E18 or E19). Schilling and Pilgrim[126] reported that the strain of rats used may impact on the optimal age to establish cultures containing VP neurons, but this may reflect slight differences in the rate of development in different strains. As a result of these studies, it is recognized that the optimal donor age for culturing magnocellular VP neurons is somewhat younger than that generally used for parvocellular systems (Table 1), but interesting differences in the characteristics of cultured VP neurons are observed even within this narrow ontogenetic window. For example, DiScala reported that E15-16 was optimal for longterm maintenance of VP neurons in culture, but in cultures established one day earlier (E14), VP gene expression is markedly dependent on the culture conditions.[114,129,145] At E14, very few neurons express VP or neurophysin unless the cultures are treated either with agents that block glucocorticoid receptors or that elevate cAMP.[114,129,145] This characteristic has proved useful for studies of VP gene expression, because the gene can be reversibly turned on and off,[114] and therefore the influence of other agents on VP gene expression can be evaluated in these cultures.[146]

Donor age as well as culture conditions also impact on the expression of OT in primary hypothalamic cultures. OT expressing neurons are not present in cultures established from E14 to E17 hypothalamus,[114,120,128,132] but a few cells can be detected in E18-19 cultures and the number of OT expressing cells can be increased by treatment with triodothyronine.[132] These findings are significant, because the VP and OT neurons are generated essentially simultaneously in the proliferative zone.[147] Therefore, at the time of plating, cells destined to differentiate into both VP and OT neurons should be present in the culture. The identification of agents that induce or inhibit VP gene expression (e.g., cAMP and glucocorticoids respectively) suggests that the lack of OT expression may also reflect the absence of inducing signals or the presence of inhibitory signals, and therefore, the dispersed hypothalamic cultures may prove useful in identifying these regulatory signals. The differential expression of VP and OT in fetal culture preparations is particularly interesting if one compares primary hypothalamic cultures with long-term slice cultures, because the long term slice cultures are characterized by OT expression, but an absence of VP expression[107] whereas in primary cultures it is OT expression that is repressed.[114,120]

As mentioned above, donor age can impact on the ability to sustain fetal neurons in culture for extended periods.[128] This is significant, because the ability to sustain fetal neurons in culture for extended periods may be important to allow differentiation and maturation of the neurons. For example, Obrietan and van den Pol[141] found that neurons in primary hypothalamic cultures established from E15 tissue underwent a developmental reversal in their Ca^{2+} responses to GABA while in culture. From days 4 to 10 of culture, neurons demonstrated increases in intracellular Ca^{2+} in response to GABA, but in older neurons (>18 days *in vitro*), GABA caused the expected decrease in intracellular Ca^{2+}. Thus, donor age is an important consideration both in establishing cultures of specific neuroendocrine neurons and in interpreting experimental results.

b.   Use of adult or neonatal tissue.   The heterogeneity of neuronal populations inherent in primary dispersed hypothalamic culture preparations can be a disadvantage for certain experiments. The presence of multiple neuronal populations impacts on interpretation of experimental results, because interactions between neuronal populations in culture has been shown, and therefore, responses to experimental manipulations can reflect either direct or indirect effects on the neuronal population of interest. Another disadvantage of the mixed population of neurons is the inability to phenotype living neurons in order to identify specific neuron types for experiments on living cells such as electrophysiological recordings or calcium imaging techniques.[135] If the cell type of interest represents a small percentage of the neurons in a culture,[114,116] picking the correct neuron for these types of measurements is like finding a needle in a haystack. Also, "dilution" of the population of interest in the total hypothalamic neuronal pool may limit detection capabilities of some end points (e.g., peptide release). One way to address these problems is to utilize microdissected hypothalamic regions from older animals in which the neurons have localized in discrete and recognizable regions of the hypothalamus. Although some investigators have used adult tissue to establish dispersed hypothalamic culture preparations,[148-150] the more severe enzymatic and mechanical treatments required to disperse mature tissue reduces the yield of neurons obtained such that issues of detection limitations still exist. Also, some investigators have still utilized the entire hypothalamus to establish these cultures.[149,150] Other investigators have capitalized on the microdissection approach to achieve preparations of selected neuroendocrine neurons,[151,152] and have even labeled the cells prior to dissection by intravenous administration of Evans Blue dye.[153] Neuroendocrine neurons are labeled by this technique due to their axonal projections to regions lacking a blood brain barrier. These acutely dissociated adult neurons have been useful for acute patch clamp recording studies,[151-153] but they do not survive long enough to re-establish neuritic processes and may not have had adequate time to re-establish a normal complement of membrane receptors following enzymatic dispersion. Therefore, their utility may be limited to specific experimental paradigms.

Neonatal or late embryonic tissue can provide the same advantage of localized tissue distribution of neurons, but may be more viable in culture. Investigators interested in evaluating the circadian function of the SCN have taken advantage of this approach. By selectively dissecting the SCN region from late embryonic or postnatal rats, they have successfully established primary cultures in which the majority of neurons can be assumed to originate from the SCN and therefore are relevant for studies of circadian properties.[154-158] These cultures have been shown to contain VP and VIP neurons,[159] to maintain circadian release of these agents,[154,156] and have been used to evaluate the electrophysiological characteristics of SCN neurons.[141,155]

3. Culture Vessel

Following dispersion, primary hypothalamic cultures have been maintained in a variety of culture vessels ranging from glass slides to large flasks to beads or membrane amenable to perifusion. The optimal culture technique varies depending upon the cell type of interest and the experimental objective. Plating density frequently impacts on neuronal survival, and most frequently a plating density that allows for establishment of a confluent glial cell layer proves optimal. However, under these conditions, neurons rapidly establish functional interactions, and thus can no longer be studied as isolated neurons. For experiments requiring isolated neurons, low density cultures have proved useful.[119] In these cultures as well as the higher density cultures the plating matrix may be critical for establishing viable cultures, and the ideal matrix may vary with the donor age.[116,126] Poly-lysine is frequently used to coat glass or plastic culture wells to achieve optimal attachment, and 3D matrixes are useful for low density cultures.[119] The size of the culture vessel is primarily determined by the experimental end point. Small culture wells are useful for studies employing morphological end points, but larger populations of cells may be required when the end point involves detection of peptide release, synthesis, or mRNA expression. Thus, larger culture dishes or flasks seeded with greater numbers of cells are frequently employed in these latter studies.[130,131,137] Unlike cultures established from cell lines (see Chapter 1), these larger cultures require more donors to provide adequate numbers of primary cells, and this can limit the utility of this approach.[130] Perifusion approaches have been devised for studying peptide release. For this purpose primary cultures have been allowed to attach to beads or capillary membranes that can subsequently be perifused directly[133] or loaded into a perifusion chamber.[139] The latter approach proved useful for studies demonstrating pulsatile secretion of LHRH.[139] Other elaborate preparations have been developed for long term recordings of spontaneous action potentials from individual SCN neurons in which cells are cultured on fixed microelecrode arrays.[158]

4. Culture Medium

The majority of primary hypothalamic cultures have been maintained in some variety of Eagle's medium supplemented either with serum (fetal bovine or horse) or in serum free medium supplemented with a variety of hormones and growth promoting agents.[160] Either Eagle's minimum essential medium (MEM) or Dulbecco's modification of MEM (DMEM) are usually used. Also, MEM or DMEM is sometimes used in a 1:1 ratio with Ham's F12 nutrient mixture. Most commonly, cultures are initially plated in medium supplemented with either fetal bovine serum or horse serum or both. In order to achieve better control over the components of the culture medium, investigators frequently switch to a "defined" medium after the culture is established. However, survival of the neurons in serum-free medium requires supplementing the medium with a variety of hormones and growth promoting agents including transferrin, putrescine, selinium, triodothyronine, corticosterone, progesterone, estradiol and arachidonic acid.[118,122,160] Thus, although the goal of using serum-free medium is to control the experimental environment, some of the components in serum must be duplicated, and some of these may impact on neuronal function of the cultures (e.g., Reference 129). Inclusion of bacitracin or other peptidase inhibitors in the culture medium markedly enhances the ability to monitor peptide release from the cultures.[161]

III. Summary

In vitro approaches have advanced the field of neuroendocrinology by providing the opportunity to study neuroendocrine systems in controlled conditions and in isolation from peripheral feedback systems. *In vitro* approaches are also beneficial, because they allow experimental manipulations and evaluation of physiological end points that are not feasible *in vivo*. As described above, a wide variety of *in vitro* approaches have been applied to neuroendocrine studies. Each approach offers advantages for certain types of studies, but the limitations of each approach must be considered when interpreting the results. Investigators must assess the relative characteristics of each approach in determining the optimal method for obtaining the specific information desired.

References

1. Sladek, C. D. and Knigge, K. M., Osmotic control of vasopressin release by organ cultured hypothalamo-neurohypophyseal explants from the rat, *Endocrinology*, 101, 1834, 1977.
2. Stern, J. E., Mitchell, T., Herzberg, V. L. and North, W. G., Secretion of vasopressin, oxytocin, and two neurophysins from rat hypothalamo-neurohypophyseal explants in organ culture, *Neuroendocrinology*, 43, 252, 1986.
3. Davis, M. D. and Kilts, C. D., Endogenous dopamine and serotonin release from the explanted rat tuberohypophyseal system: Effects of electrical stimulation and neurotensin, *Life Sci.*, 40, 1869, 1987.
4. Clough, R. W., Hoffman, G. E. and Sladek, C. D., Peripuberal development of noradrenergic stimulation of luteinizing hormone-releasing hormone neurosecretion *in vitro*, *Brain Res.*, 446, 121, 1988.
5. Calogero, A. E., Gallucci, W. T., Kling, M. A., Chrousos, G. P. and Gold, P. W., Cocaine stimulates rat hypothalamic corticotropin-releasing hormone secretion *in vitro*, *Brain Res.*, 505, 7, 1989.
6. Sweep, C. G. J. and Wiegant, V. M., Release of β-endorphin-immunoreactivity from rat pituitary and hypothalamus *in vitro*: Effects of isoproterenol, dopamine, corticotropin-releasing factor and arginine[8]-vasopressin, *Biochem. Biophys. Res. Commun.*, 161, 221, 1989.
7. Hillhouse, E. W. and Milton, N. G. N., Effect of noradrenaline and gamma-aminobutyric acid on the secretion of corticotrophin-releasing factor-41 and arginine vasopressin from the rat hypothalamus *in vitro*, *J. Endocrinol.*, 122, 719, 1989.

8. Ohgo, S., Nakatsuru, K., Ishikawa, E. and Matsukura, S., Stimulation of cholecysto-kinin (CCK) release from superfused rat hypothalamo-neurohypophyseal complexes by interleukin-1 (IL-1), *Brain Res.*, 593, 25, 1992.

9. Garris, P. A. and Ben-Jonathan, N., A compartmentalized chamber for studying dopamine neurons in an hypothalamo-pituitary explant, *J. Neurosci. Methods*, 49, 113, 1993.

10. Yasin, S. A., Costa, A., Forsling, M. L. and Grossman, A., Interleukin-1β and interleukin-6 stimulate neurohypophysial hormone release *in vitro*, *J. Neuroendocrinol.*, 6, 179, 1994.

11. Bernardini, R., Chiarenza, A., Kamilaris, T. C., Renaud, N., Lempereur, L., Demitrack, M., Gold, P. W. and Chrousos, G. P., *In vivo* and *in vitro* effects of arginine-vasopressin receptor antagonists on the hypothalamic-pituitary-adrenal axis in the rat, *Neuroendocrinology*, 60, 503, 1994.

12. Scarceriaux, V., Pelaprat, D., Forgez, P., Lhiaubet, A. M. and Rostene, W., Effects of dexamethasone and forskolin on neurotensin production in rat hypothalamic cultures, *Endocrinology*, 136, 2554, 1995.

13. Eggena, P. and Thorn, N. A., Vasopressin release from the rat supraoptico-neurohypophysial system *in vitro* in response to hypertonicity and acetylcholine, *Acta Endocrinol.*, 65, 441, 1970.

14. Nordmann, J. J., Bianchi, R. E., Dreifuss, J. -J. and Ruf, K. B., Release of posterior pituitary hormones from the entire hypothalamo-neurohypophysial system *in vitro*, *Brain Res.*, 25, 669, 1971.

15. Sladek, C. D. and Joynt, R. J., Cholinergic involvement in osmotic control of vasopressin release by the organ cultured rat hypothalamo-neurohypophyseal system, *Endocrinology*, 105, 367, 1979.

16. Ishikawa, S. -E., Saito, T. and Yoshida, S., The effect of prostaglandins on the release of arginine vasopressin from the guinea pig hypothalamo-neurohypophyseal complex in organ culture, *Endocrinology*, 108, 193, 1981.

17. Sladek, C. D., Blair, M. L. and Ramsay, D. J., Further studies on the role of angiotensin in the osmotic control of vasopressin release by the organ cultured rat hypothalamo-neurohypophyseal system, *Endocrinology*, 111, 599, 1982.

18. Ishikawa, S. E. and Schrier, R. W., Role of calcium in osmotic and nonosmotic release of vasopressin from rat organ culture, *Am. J. Physiol.*, 244, R703, 1983.

19. Gregg, C. M. and Sladek, C. D., A compartmentalized, organ-cultured hypothalamo-neurohypophysial system for the study of vasopressin release, *Neuroendocrinology*, 38, 397, 1984.

20. Sladek, C. D. and Armstrong, W. E., GABA antagonists stimulate vasopressin release from organ-cultured hypothalamo-neurohypophyseal explants, *Endocrinology*, 120, 1576, 1987.

21. Schiavone, M. T., Santos, R. A. S., Brosnihan, K. B., Khosla, M. C. and Ferrario, C. M., Release of vasopressin from the rat hypothalamo-neurohypophysial system by angiotensin-(1-7) heptapeptide, *Proc. Natl. Acad. Sci. USA*, 85, 4095, 1988.

22. Rossi, N. F. and Schrier, R. W., Anti-calmodulin agents affect osmotic and angiotensin II-induced vasopressin release, *Am. J. Physiol.*, 256, E516, 1989.

23. Yagil, C. and Sladek, C. D., Osmotic regulation of vasopressin and oxytocin release is rate sensitive in hypothalamoneurohypophysial explants, *Am. J. Physiol.*, 258, R492, 1990.

24. Meeker, R. B., Michels, K. M. and Hayward, J. N., Vasopressin and oxytocin regulation of cyclic AMP accumulation in rat hypothalamo-neurohypophysial explants *in vitro*, *Neurosci. Lett.*, 114, 225, 1990.

25. Michels, K. M., Meeker, R. B. and Hayward, J. N., Muscarinic cholinergic control of vasopressin secretion from the acute hypothalamoneurohypophysial explant, *Neuroendocrinology*, 54, 219, 1991.

26. Ohgo, S., Nakatsuru, K., Ishikawa, E. and Matsukura, S., Interleukin-1 (IL-1) stimulates the release of corticotropin-releasing factor (CRF) from superfused rat hypothalamo-neurohypophyseal complexes (HNC) independently of the histaminergic mechanism, *Brain Res.*, 558, 217, 1991.

27. Rossi, N. F., Effect of endothelin-3 on vasopressin release *in vitro* and water excretion *in vivo* in Long-Evans rats, *J. Physiol. (London)*, 461, 501, 1993.

28. Rossi, N. F., Cation channel mechanisms in ET-3-induced vasopressin secretion by rat hypothalamo-neurohypophysial explants, *Am. J. Physiol.*, 268, E467, 1995.

29. Papanek, P. E., Sladek, C. D. and Raff, H., Corticosterone inhibition of osmotically-stimulated vasopressin release from hypothalamic-neurohypophysial explants, *Am. J. Physiol.*, 272, R158, 1997.

30. Pearson, D., Shainberg, A., Osinchak, J. and Sachs, H., The hypothalamo-neurohypophysial complex in organ culture: morphological and biochemical characteristics, *Endocrinology*, 96, 982, 1974.

31. Van Tol, H. H. M., Promes, L. W., Kiss, J. Z. and Burbach, J. P. H., Osmotic regulation of vasopressin and oxytocin gene expression in the hypothalamus, in *Recent Progress in Posterior Pituitary Hormones 1988*, S. Yoshida and L. Share, Eds., Elsevier Science, Amsterdam, 1988, 241.

32. Yagil, C. and Sladek, C. D., Effect of extended exposure to hypertonicity on vasopressin mRNA content in hypothalamo-neurohypophyseal explants, *Endocrinology*, 127, 1428, 1990.

33. Sladek, C. D., Fisher, K. Y., Sidorowicz, H. E. and Mathiasen, J. R., Osmotic stimulation of vasopressin mRNA content in the supraoptic nucleus requires synaptic activation, *Am. J. Physiol.*, 268, R1034, 1995.

34. Swenson, K. L. and Sladek, C. D., Gonadal steroid modulation of vasopressin secretion in response to osmotic stimulation, *Endocrinology*, 138, 2089, 1997.

35. Armstrong, W. E. and Sladek, C. D., Spontaneous "phasic-firing" in supraoptic neurons recorded from hypothalamo-neurohypophyseal explants *in vitro*, *Neuroendocrinology*, 34, 405, 1982.

36. Bourque, C. W. and Renaud, L. P., Activity patterns and osmosensitivity of rat supraoptic neurones in perfused hypothalamic explants, *J. Physiol. (London)*, 346, 631, 1984.

37. Randle, J. C. R., Bourque, C. W. and Renaud, L. P., Characterization of spontaneous and evoked inhibitory postsynaptic potentials in rat supraoptic neurosecretory neurons *in vitro*, *J. Neurophys.*, 56, 1703, 1986.

38. Hatton, G. I., Emerging concepts of structure-function dynamics in adult brain: The hypothalamo-neurohypophysial system, *Prog. Neurobiol.*, 34, 437, 1990.

39. Sladek, C. D. and Johnson, A. K., The effect of anteroventral third ventricle lesions on vasopressin release by organ cultured hypothalamo-neurohypophyseal explants, *Neuroendocrinology*, 37, 78, 1983.

40. Nissen, R., Bourque, C. W. and Renaud, L. P., Membrane properties of organum vasculosum lamina terminalis neurons recorded *in vitro*, *Am. J. Physiol.*, 264, R811, 1993.

41. Bjorklund, A., Moore, R. Y., Nobin, A. and Stenevi, U., The organization of tuberohypophyseal and reticulo-infundibular catecholamine neuron systems in the rat brain, *Brain Res.*, 51, 171, 1973.

42. Khachaturian, H., Lewis, M. E., Schafer, M. K. -H. and Watson, S. J., Anatomy of the CNS opioid systems, *TINS*, 8, 111, 1985.

43. Gregg, C. M., The compartmentalized hypothalamo-neurohypophysial system: Evidence for a neurohypophysial action of acetylcholine on vasopressin release, *Neuroendocrinology*, 40, 423, 1985.

44. Rossi, N. F. and Brooks, D. P., κ-Opioid agonist inhibition of osmotically induced AVP release: preferential action at hypothalamic sites, *Am. J. Physiol.*, 270, E367, 1996.

45. Bicknell, R. J. and Leng, G., Endogenous opiates regulate oxytocin but not vasopressin secretion from the neurohypophysis, *Nature*, 298, 161, 1982.

46. Bondy, C. A., Gainer, H. and Russell, J. T., Dynorphin A inhibits and naloxone increases the electrically stimulated release of oxytocin but not vasopressin from the terminals of the neural lobe, *Endocrinology*, 122, 1321, 1988.

47. Bondy, C. A., Jensen, R. T., Brady, L. S. and Gainer, H., Cholecystokinin evokes secretion of oxytocin and vasopressin from rat neural lobe independent of external calcium, *Proc. Natl. Acad. Sci. USA*, 86, 5198, 1989.

48. Russell, J. A., Coombes, J. E., Leng, G. and Bicknell, R. J., Morphine tolerance and inhibition of oxytocin secretion by kappa-opioids acting on the rat neurohypophysis, *J. Physiol. (London)*, 469, 365, 1993.

49. Leng, G., Bicknell, R. J., Brown, D., Bowden, C., Chapman, C. and Russell, J. A., Stimulus-induced depletion of pro-enkephalins, oxytocin and vasopressin and pro-enkephalin interaction with posterior pituitary hormone release *in vitro*, *Neuroendocrinology*, 60, 559, 1994.

50. Bicknell, R. J., Brown, D., Chapman, C., Hancock, P. D. and Leng, G., Reversible fatigue of stimulus-secretion coupling in the rat neurohypophysis, *J. Physiol.*, 348, 601, 1984.

51. Shaw, F. D., Bicknell, R. J. and Dyball, R. E. J., Facilitation of vasopressin release from the neurohypophysis by application of electrical stimuli in bursts, *Neuroendocrinology*, 39, 371, 1984.

52. Bondy, C. A., Gainer, H. and Russell, J. T., Effects of stimulus frequency and potassium channel blockade on the secretion of vasopressin and oxytocin from the neurohypophysis, *Neuroendocrinology*, 46, 258, 1987.

53. Costa, A., Yasin, S. A., Hucks, D., Forsling, M. L., Besser, G. M. and Grossman, A., Differential effects of neuroexcitatory amino acids on corticotropin-releasing hormone-41 and vasopressin release from rat hypothalamic explants, *Endocrinology*, 131, 2595, 1992.

54. Raber, J., Pich, E. M., Koob, G. F. and Bloom, F. E., IL-1 beta potentiates the acetylcholine-induced release of vasopressin from the hypothalamus *in vitro*, but not from the amygdala, *Neuroendocrinology*, 59, 208, 1994.

55. Uckimua, D., Katafuchi, T., Hori, T. and Yanaihara, N., Facilitatory effect of pituitary adenylate cyclase activating polypeptide (PACAP) on neurons in the magnocellular portion of the rat hypothalamic paraventricular nucleus (PVN) *in vitro*, *J. Neuroendocrinol.*, 8, 137, 1996.

56. Moos, F., Freund-Mercier, M. J., Guerne, Y., Guerne, J. M., Stoeckel, M. E. and Richard, P., Release of oxytocin and vasopressin by magnocellular nuclei *in vitro*: specific facilitatory effect of oxytocin on its own release, *J. Endocrinol.*, 102, 63, 1984.

57. Mason, W. T., Hatton, G. I., Ho, Y. W., Chapman, C. and Robinson, I. C., Central release of oxytocin, vasopressin and neurophysin by magnocellular neurone depolarization: evidence in slices of guinea pig and rat hypothalamus, *Neuroendocrinology*, 42, 311, 1986.

58. Landgraf, R. and Ludwig, M., Vasopressin release within the supraoptic and paraventricular nuclei of the rat brain: Osmotic stimulation via microdialysis, *Brain Res.*, 558, 191, 1991.

59. Ota, M., Crofton, J. T., Festavan, G. and Share, L., Central carbachol stimulates vasopressin release into interstitial fluid adjacent to the paraventricular nucleus, *Brain Res.*, 592, 249, 1992.

60. Neumann, I., Ludwig, M., Engelmann, M., Pittman, Q. J. and Landgraf, R., Simultaneous microdialysis in blood and brain: Oxytocin and vasopressin release in response to central and peripheral osmotic stimulation and suckling in the rat, *Neuroendocrinology*, 58, 637, 1993.

61. Sladek, C. D., Fisher, K. Y., Sidorowicz, H. E. and Mathiasen, J. R., cAMP regulation of vasopressin mRNA content in hypothalamo-neurohypophyseal explants, *Am. J. Physiol.*, 271, R554, 1996.

62. Bourque, C. W. and Renaud, L. P., *In vitro* neurophysiology of identified rat hypothalamic 'neuroendocrine' neurons, *Neuroendocrinology*, 36, 161, 1983.

63. Armstrong, W. E., Gallagher, M. J. and Sladek, C. D., Noradrenergic stimulation of supraoptic neuronal activity and vasopressin release *in vitro*: Mediation by a alpha-1-receptor, *Brain Res.*, 365, 192, 1986.

64. Bourque, C. W., Randle, J. C. R. and Renaud, L. P., Non-synaptic depolarizing potentials in rat supraoptic neurones recorded *in vitro*, *J. Physiol. (London)*, 376, 493, 1986.

65. Bourque, C. W., Ionic basis for the intrisic activation of rat supraoptic neurones by hyperosmotic stimuli, *J. Physiol.*, 417, 263, 1989.

66. Smith, B. N. and Armstrong, W. E., Tuberal supraoptic neurons — I. Morphological and electrophysiological characteristics observed with intracellular recording and biocytin filling *in vitro*, *Neuroscience*, 38, 469, 1990.

67. Hu, B. and Bourque, C. W., NMDA receptor-mediated rhythmic bursting activity in rat supraoptic nucleus neurones *in vitro*, *J. Physiol. (London)*, 458, 667, 1992.

68. Armstrong, W. E., Wilson, C. J., Gallagher, M. J. and Sladek, C. D., Quantitative comparisons between the electrical activity of supraoptic neurons and vasopressin release *in vitro*, *J. Neuroendocrinol.*, 1, 215, 1989.

69. Randle, J. C. R., Bourque, C. W. and Renaud, L. P., α1-adrenergic receptor activation depolarizes rat supraoptic neurosecretory neurons *in vitro*, *Am. J. Physiol.*, 251, R569, 1986.

70. Armstrong, W. E. and Sladek, C. D., Evidence for excitatory actions of histamine on supraoptic neurons *in vitro*: Mediation by a H1-type receptor, *Neuroscience*, 16, 307, 1985.

71. Richard, D. and Bourque, C. W., Synaptic activation of rat supraoptic neurons by osmotic stimulation of the organum vasculosum lamina terminalis, *Neuroendocrinology*, 55, 609, 1992.

72. Earnest, D. J. and Sladek, C. D., Model for studying circadian mechanisms: Vasopressin rhythms in suprachiasmatic cultures, *Brain Res.*, 382, 139, 1986.

73. Earnest, D. J. and Sladek, C. D., Circadian vasopressin release from perifused rat suprachiasmatic explants *in vitro*: Effects of acute stimulation, *Brain Res.*, 422, 398, 1987.

74. Earnest, D. J., Digiorgio, S. M. and Sladek, C. D., Effects of tetrodotoxin on the circadian pacemaker mechanism in suprachiasmatic explants *in vitro*, *Brain Res. Bull.*, 26, 677, 1991.

75. Wray, S., Castel, M. and Gainer, H., Characterization of the suprachiasmatic nucleus in organotypic slice explant cultures, *Microsc. Res. Tech.*, 25, 46, 1993.

76. Gillette, M. U. and Prosser, R. A., Circadian rhythm of the rat suprachiasmatic brain slice is rapidly reset by daytime application of cAMP analogs, *Brain Res.*, 474, 348, 1988.

77. Bos, N. P. A. and Mirmiran, M., Circadian rhythms in spontaneous neuronal discharges of the cultured suprachiasmatic nucleus, *Brain Res.*, 511, 158, 1990.

78. Liou, S. Y., Shibata, S., Albers, H. E. and Ueki, S., Effects of GABA and anxiolytics on the single unit discharge of suprachiasmatic neurons in rat hypothalamic slices, *Brain Res. Bull.*, 25, 103, 1990.

79. Medanic, M. and Gillette, M. U., Serotonin regulates the phase of the rat suprachiasmatic circadian pacemaker *in vitro* only during the subjective day, *J. Physiol. (London)*, 450, 629, 1992.

80. Prosser, R. A. and Gillette, M. U., The mammalian circadian clock in the suprachiasmatic nuclei is reset *in vitro* by cAMP, *J. Neurosci.*, 9, 1073, 1989.

81. Hartter, D. E. and Ramirez, V. D., Responsiveness of immature vs. adult male rat hypothalami to dibutyryl cyclic AMP- and forskolin-induced LHRH release *in vitro*, *Neuroendocrinology*, 40, 476, 1985.

82. Bourguignon, J. P., Gerard, A., Mathieu, J., Mathieu, A. and Franchimont, P., Maturation of the hypothalamic control of pulsatile gonadotropin-releasing hormone secretion at onset of puberty. I. Increased activation of N-methyl-D-aspartate receptors, *Endocrinology*, 127, 873, 1990.

83. Bourguignon, J. P., Gerard, A. and Franchimont, P., Maturation of the hypothalamic control of pulsatile gonadotropin-releasing hormone secretion at onset of puberty: II. Reduced potency of an inhibitory autofeedback, *Endocrinology*, 127, 2884, 1990.

84. Kim, K., Lee, C. S., Cho, W. K. and Ramirez, V. D., Effect of chronic administration of progesterone on the naloxone-induced LHRH release from hypothalmi of ovariectomized, estradiol-primed prepubertal rats, *Life Sci.*, 43, 609, 1988.

85. Clough, R. W., Hoffman, G. E. and Sladek, C. D., Synergistic interaction between opioid receptor blockade and α-adrenergic stimulation on luteinizing hormone-releasing hormone (LHRH) secretion *in vitro*, *Neuroendocrinology*, 51, 131, 1990.

86. Calagero, A. E., Burrello, N., Ossino, A. M., Weber, R. F. and D'Agata, R., Interaction between prolactin and catecholamines on hypothalmic GnRH release *in vitro*, *J. Endocrinol.*, 151, 269, 1996.

87. Patchev, V. K., Karalis, K. and Chrousos, G. P., Effects of excitatory amino acid transmitters on hypothalamic corticotropin-releasing hormone (CRH) and arginine-vasopressin (AVP) release *in vitro*: Implications in pituitary-adrenal regulation, *Brain Res.*, 633, 312, 1994.

88. Hillhouse, E. W., Interleukin-2 stimulates the secretion of arginine vasopressin but not corticotropin-releasing hormone from rat hypothalamic cells *in vitro*, *Brain Res.*, 650, 323, 1994.

89. Weber, R. F. and Calogero, A. E., Prolactin stimulates rat hypothalamic corticotropin releasing hormone and pituitary adrenocorticotropin secretion *in vitro*, *Neuroendocrinology*, 54, 248, 1991.

90. Hoffman, G. E., Phelps, C. J., Khachaturian, H. and Sladek, J. R.,Jr., Neuroendocrine projections to the median eminence, *Current Topics Neuroendo.*, 7, 161, 1986.

91. Lewis, B. M., Ismail, I. S., Issa, B., Peters, J. R. and Scanlon, M. F., Desensitisation of somatostatin, TRH and GHRH responses to glucose in the diabetic (Goto-Kakizaki) rat hypothalamus, *J. Endocrinol.*, 151, 13, 1996.

92. Aguila, M. C. and McCann, S. M., Growth hormone increases somatostatin release and messenger ribonucleic acid levels in the rat hypothalamus, *Brain Res.*, 623, 89, 1993.

93. Aguila, M. C., Boggaram, V. and McCann, S. M., Insulin-like growth factor I modulates hypothalamic somatostatin through a growth hormone releasing factor increased somatostatin release and messenger ribonucleic acid levels, *Brain Res.*, 625, 213, 1993.

94. Honegger, J., Spagnoli, A., D'urso, R., Navarra, P., Tsagarakis, S., Beessr, G. M. and Grossman, A. B., Interleukin-1β modulates the acute release of growth hormone-releasing hormone and somatostatin from rat hypothalamus *in vitro*, whereas tumor necrosis factor and interleukin-6 have no effect, *Endocrinology*, 129, 1275, 1991.

95. Davis, M. D., Haas, H. L. and Lichtensteiger, W., The hypothalamohypophyseal system *in vitro*: Electrophysiology of the pars intermedia and evidence for both excitatory and inhibitory inputs, *Brain Res*, 334, 97, 1985.

96. Dreiffus, J. J. and Gahwiler, B. H., Hypothalamic neurons in culture. I. A short review of the literature, *J. Physiol. Paris.*, 75, 15, 1979.

97. Sakai, K. K., Marks, B. H., George, J. M. and Koestner, A., The isolated organ cultured supraoptic nucleus as a neuropharmacological test system, *J. Pharmacol. Exp. Ther.*, 190, 482, 1974.

98. Gahwiler, B. H., Sandoz, P. and Dreifuss, J. J., Neurones with synchronous bursting discharges in organ cultures of the hypothalamic supraoptic nucleus area, *Brain Res*, 151, 245, 1978.

99. Bennett, B. A., Clodfelter, J., Sundberg, D. K. and Morris, M., Cultured hypothalamic explants from spontaneously hypertensive rats have decreased vasopressin and oxytocin content and release, *Am. J. Hypertens.*, 2, 46, 1989.

100. Gahwiler, B. H., Organotypic cultures of neural tissue, *TINS*, 11, 484, 1988.

101. Toran-Allerand, C. D., The luteinizing hormone-releasing hormone (LHRH) neuron in cultures of the newborn mouse hypothalamus/preoptic area:ontogenic aspects and response to steroid, *Brain Res*, 149, 257, 1978.

102. Marson, A. -M. and Privat, A., *In vitro* differentiation of hypothalamic magnocellular neurons of guinea pigs, *Cell Tiss. Res.*, 203, 393, 1979.

103. Gahwiler, B. H., Slice cultures of cerebellar, hippocampal, and hypothalamic tissue, *Experientia*, 40, 235, 1984.

104. Gainer, H., Kusano, K. and Wray, S., Hypothalamic slice-explant cultures as models for the long-term study of gene expression and cellular activity, *Regul. Pept.*, 45, 25, 1993.

105. Tominaga, K., Geusz, M. E., Michel, S. and Inouye, S. T., Calcium imaging in organotypic cultures of the rat suprachiasmatic nucleus, *Neuroreport*, 51, 1901, 1994.

106. Wray, S., Gahwiler, B. H. and Gainer, H., Slice cultures of LHRH neurons in the presence and absence of brainstem and pituitary, *Peptides*, 9, 1151, 1988.

107. Wray, S., Kusano, K. and Gainer, H., Maintenance of LHRH and oxytocin neurons in slice explants cultured in serum-free media: Effects of tetrodotoxin on gene expression, *Neuroendocrinology*, 54, 327, 1991.

108. Szafarczyk, A., Feuvrier, E., Siaud, P., Rondouin, G., Lacoste, M., Gaillet, S., Malaval, F. and Assenmacher, I., Removal of adrenal steroids from the medium reverses the stimulating effect of catecholamines on corticotropin-releasing hormone neurons in organotypic cultures, *Neuroendocrinology*, 61, 517, 1995.

109. Belenky, M., Wagner, S., Yarom, Y., Matzner, H., Cohen, S. and Castel, M., The suprachiasmatic nucleus in stationary organotypic culture, *Neuroscience*, 70, 127, 1996.

110. Shinohara, K., Honma, S., Katsuno, Y., Abe, H. and Honma, K., Circadian rhythms in the release of vasoactive intestinal polypeptide and arginine-vasopressin in organotypic slice culture of rat suprachiasmatic nucleus, *Neurosci. Lett.*, 170, 183, 1994.

111. Tominaga, K., Inouye, S. I. and Okamura, H., Organotypic slice culture of the rat suprachiasmatic nucleus: sustenance of cellular architecture and circadian rhythm, *Neuroscience*, 59, 1025, 1994.

112. Shinohara, K. and Oka, T., Protein synthesis inhibitor phase shifts vasopressin rhythms in long-term suprachiasmatic cultures, *Neuroreport*, 5, 2201, 1994.

113. Shinohara, K., Honma, S., Katsuno, Y., Abe, H. and Honma, K., Two distinct oscillators in the rat suprachiasmatic nucleus *in vitro*, *Proc. Natl. Acad. Sci. USA*, 92, 7396, 1995.

114. Sladek, C. D. and Gallagher, M. J., The stimulation of vasopressin gene expression in cultured hypothalamic neurons by cyclic adenosine 3',5'-monophosphate is reversible, *Endocrinology*, 133, 1320, 1993.

115. Wahle, P., Müller, T. H. and Swandulla, D., Characterization of neurochemical phenotypes in cultured hypothalamic neurons with immunohistochemistry and *in situ* hybridization, *Brain Res.*, 611, 37, 1993.

116. Jirikowski, G., Reisert, I. and Pilgrim, Ch., Neuropeptides in dissociated cultures of hypothalamus and septum: Quantitation of immunoreactive neurons, *Neuroscience*, 6, 1953, 1981.

117. Clarke, M. J., Lowry, P. and Gillies, G., Assessment of corticotropin-releasing factor, vasopressin and somatostatin secretion by fetal hypothalamic neurons in culture, *Neuroendocrinology*, 46, 147, 1987.

118. Clarke, M. J. and Gillies, G. E., Comparison of peptide release from fetal rat hypothalmic neurones cultured in defined media and serum-containing media, *J. Endocrinol.*, 116, 349, 1988.

119. Dudley, C. A., Coates, P. W. and Moss, R. L., Solitary hypothalamic neurons inherently express vasopressin and tyrosine hydroxylase, *Peptides*, 10, 1205, 1989.

120. Di Scala-Guenot, D., Strosser, M. T., Felix, J. M. and Richard, P., Expression of vasopressin and opiates but not of oxytocin genes studied by *in situ* hybridization in embryonic rat brain primary cultures, *Dev. Brain Res.*, 56, 35, 1990.

121. de los Frailes, M. T., Cacicedo, L., Fernandez, G., Tolon, R. M., Jesus Lorenzo, M., Aguado, F. and Sanchez Franco, F., Role of locally produced growth hormone-releasing factor in somatostatin regulation by fetal rat brain cells in culture, *Neuroendocrinology*, 55, 221, 1992.

122. Gillies, G. and Davidson, K., GABAergic influences on somatostatin secretion from hypothalamic neurons cultured in defined medium, *Neuroendocrinology*, 55, 248, 1992.
123. Rage, F., Benyassi, A., Arancibia, S. and Tapia-Arancibia, L., Gamma-aminobutyric acid-glutamate interaction in the control of somatostatin release from hypothalamic neurons in primary culture: *in vivo* corroboration, *Endocrinology*, 130, 1056, 1992.
124. Denizeau, F., Dube, D., Antakly, T., Lemay, A., Parent, A., Pelletier, G. and Labrie, F., Attempts to demonstrate peptide localization and secretion in primary cell cultures of fetal rat hypothalamus, *Neuroendocrinology*, 32, 96, 1981.
125. Theodosis, D. T., Legendre, P., Vincent, J. D. and Cooke, I., Immunocytochemically identified vasopressin neurons in culture show slow, calcium-dependent electrical responses, *Science*, 221, 1052, 1981.
126. Schilling, K. and Pilgrim, C., Hypothalamo-neurohypophysial neurons *in vitro*: Developmental potentials depend on the donor rat stock, *J. Neurosci. Res.*, 18, 432, 1987.
127. Schilling, K. L. and Pilgrim, C., Developmental effect of dimethylsulfoxide on hypothalamo-neurohypophysial neurons, *J. Neurosci. Res.*, 19, 27, 1988.
128. DiScala-Guenot, D., Strosser, M. T., Sarlière, L. L., Legros, J. J. and Richard, P., Development of neurophysin-containing neurons in primary cultures of rat hypothalami is related to the age of the embryo: Morphological study and comparison of *in vivo* and *in vitro* neurophysins, oxytocin, and vasopressin content, *J. Neurosci. Res.*, 25, 94, 1990.
129. Oeding, P., Schilling, K. and Schmale, H., Vasopressin expression in cultured neurons is stimulated by cyclic AMP, *J. Neuroendocrinol.*, 2, 859, 1990.
130. Emanuel, R. L., Girard, D. M., Thull, D. L. and Majzoub, J. A., Regulated expression of vasopressin gene by cAMP and phorbol ester in primary rat fetal hypothalamic cultures, *Mol. Cell. Endocrinol.*, 86, 29, 1992.
131. Hu, S. -B., Tannahill, L. A. and Lightman, S. L., Regulation of arginine vasopressin mRNA in rat fetal hypothalamic cell culture. Role of protein kinases and glucocorticoids, *J. Mol. Endocrinol.*, 10, 51, 1993.
132. Madarasz, E., Kornyei, Z., Poulain, D. A. and Theodosis, D. T., Development of oxytocinergic neurons in monolayer cultures derived from embryonic, fetal and postnatal rat hypothalami, *J. Neuroendocrinol.*, 4, 433, 1992.
133. Scanlon, M. F., Robbins, R. J., Bolaffi, J. L., Jackson, I. M. D. and Reichlin, S., Characterisation of somatostatin and TRH release by rat hypothalamic and cerebral cortical neurons maintained on a capillary membrane perfusion system, *Neuroendocrinology*, 37, 269, 1983.
134. Heidet, V., Faivre-Bauman, A., Kordon, C., Loudes, C., Rasolonjanahary, S. and Epelbaum, J., Functional maturation of somatostatin neurons and somatostatin receptors during development of mouse hypothalamus *in vivo* and *in vitro*, *Brain Res. Dev. Brain Res.*, 57, 85, 1990.
135. Desarmenien, M. G., Devic, E., Rage, F., Dayanithi, G., Tapia-Arancibia, L. and Richard, P., Synchronous development of spontaneous and evoked calcium-dependent properties in hypothalamic neurons, *Brain Res. Dev. Brain Res.*, 79, 85, 1994.
136. Fernandez-Vazquez, G., Cacicedo, L., Lorenzo, M. J., Tolon, R., Lopez, J. and Sanchez-Franco, F., Corticosterone modulates growth hormone-releasing factor and somotostatin in fetal rat hypothalamic cultures, *Neuroendocrinology*, 61, 31, 1995.

137. Emanuel, R. L., Girard, D. M., Thull, D. L. and Majzoub, J. A., Second messengers involved in the regulation of corticotropin-releasing hormone mRNA and peptide in cultured rat fetal hypothalamic primary cultures, *Endocrinology*, 126, 3016, 1990.

138. Ackland, J. F., Nikolics, K., Seeburg, P. H. and Jackson, I. M., Molecular forms of gonadotropin-releasing hormone associated peptide (GAP): changes within the rat hypothalamus and release from hypothalamic cells *in vitro*, *Neuroendocrinology*, 48, 376, 1988.

139. Krsmanovic, L. Z., Stojilkovic, S. S., Meerelli, F., Dufour, S. M., Virmani, M. A. and Catt, K. J., Calcium signaling and episodic secretion of gonadotropin-releasing hormone in hypothalamic neurons, *Proc. Natl. Acad. Sci. USA*, 89, 8462, 1992.

140. Porter, J. C., Kedzierski, W., Aguila-Mansilla, N. and Jorquera, B. A., Expression of tyrosine hydoxylase in cultured brain cells: Stimulation with an extractable pituitary cytotropic factor, *Endocrinology*, 126, 2474, 1990.

141. Obrietan, K. and Van den Pol, A. N., GABA neurotransmission in the hypothalamus: developmental reversal from Ca^{2+} elevating to depressing, *J. Neurosci.*, 15, 5065, 1995.

142. Kapcala, L. P., Juang, H. H. and Weng, C. F., Corticotropin-releasing hormone and dexamethasone do not alter secretion of immunoreactive beta-endorphin from dissociated fetal hypothalamic cell cultures, *Brain Res.*, 532, 76, 1990.

143. Yang, Z. and Lim, A. T., Progesterone, but not estrogen, modulates, the cAMP system mediated ir-β-endorphin secretion and POMC mRNA expression from rat hypothalamic cells in culture, *Brain Res*, 678, 251, 1995.

144. Huang, W., Coi, C. L., Yang, Z., Copolov, D. L. and Lim, A. T., Forskolin-induced immunoreactive atrial natriuretic peptide (ANP) secretion and pro-ANP messenger ribonucleic acid expression of hypothalamic neurons in culture: Modulation by glucocorticoids, *Endocrinology*, 128, 2591, 1991.

145. Schilling, K., Schmale, H., Oeding, P. and Pilgrim, C., Regulation of vasopressin expression in cultured diencephalic neurons by glucocorticoids, *Neuroendocrinology*, 53, 528, 1991.

146. Mathiasen, J. R., Larson, E. R., Ariano, M. A. and Sladek, C. D., Neurophysin expression is stimulated by dopamine D_1 agonist in dispersed hypothalamic cultures, *Am. J. Physiol.*, 270, R404, 1996.

147. Altman, J. and Bayer, S. A., Development of the diencephalon in the rat. II. Correlation of the embryonic development of the hypothalamus with the time of origin of its neurons, *J. Comp. Neurol.*, 182, 973, 1978.

148. Sang U, H., Erickson, G. F. and Watkins, W. B., Long term primary monolayer culture of adult murine magnocellular neurons, *Endocrinology*, 108, 1810, 1981.

149. Richardson, S. B., Greenleaf, P. and Hollander, C. S., Somatostatin release from dispersed hypothalmic cells — Effects of membrane depolarization, calcium and glucose deprivation, *Brain Res*, 266, 75, 1983.

150. Yamaguchi, M., Yoshimoto, Y., Komua, H., Koike, K., Matsuzaki, N., Hirota, K., Miyake, A. and Tanizawa, O., Interleukin 1β and tumour necrosis factor alpha stimulate the release of gonadotropin-releasing hormone and interleukin 6 by primary cultured rat hypothalamic cells, *Acta Endocrinol. (Copenhagen)*, 123, 476, 1990.

151. Cobbett, P. and Weiss, M. L., Voltage clamp recordings from identified dissociated neuroendocrine cells of the adult supraoptic nucleus, *J. Neuroendocrinol.*, 2, 267, 1990.

152. Oliet, S. H. and Bourque, C. W., Properties of supraoptic magnocellular neurones isolated from the adult rat, *J. Physiol. (London)*, 455, 291, 1992.
153. Weiss, M. L. and Cobbett, P., Intravenous injection of Evans Blue labels magnocellular neuroendocrine cells of the rat supraoptic nucleus *in situ* and after dissociation, *Neuroscience*, 48, 383, 1992.
154. Murakami, N., Takamure, M., Takahashi, K., Utunomiya, K., Kuroda, H. and Etoh, T., Long-term cultured neurons from rat suprachiasmatic nucleus retain the capacity for circadian oscillation of vasopressin release, *Brain Res.*, 545, 347, 1991.
155. Walsh, I. B., Van den Berg, R. J., Marani, E. and Rietveld, W. J., Spontaneous and stimulated firing in cultured rat suprachiasmatic neurons, *Brain Res.*, 588, 120, 1992.
156. Watanabe, K., Koibuchi, N., Ohtake, H. and Yamaoka, S., Circadian rhythms of vasopressin release in primary cultures of rat suprachiasmatic nucleus, *Brain Res.*, 624, 115, 1993.
157. Kawahara, F., Saito, H. and Katsuki, H., Primary culture of postnatal rat suprachiasmatic neurons in serum-free supplemented medium, *Brain Res.*, 651, 101, 1994.
158. Welsh, D. K., Logothetis, D. E., Meister, M. and Reppert, S. M., Individual neurons dissociated from rat suprachiasmatic nucleus express independently phased circadian firing rhythms, *Neuron*, 14, 697, 1995.
159. Walsh, I., Marani, E., v.d.Berg, R. J. and Rietveld, W. J., The suprachiasmatic nucleus of the rat hypothalamus in culture: an anatomical and electrophysiological study, *Eur. J. Morphol.*, 28, 317, 1990.
160. Torres-Aleman, I., Naftolin, F. and Robbins, R. J., Trophic effects of insulin-like growth factor-I on fetal rat hypothalamic cells in culture, *Neuroscience*, 35, 601, 1990.
161. McKelvy, J. F., LeBlanc, P., Laudes, C., Perrie, S., Grimm-Jorgensen, Y. and Kordon, C., The use of bacitracin as an inhibitor of the degradation of thyrotropin releasing factor and luteinizing hormone releasing factor, *Biochem. Biophys. Res. Commun.*, 73, 507, 1976.

Chapter 6

Intracellular Recording from Hypothalamic Cells that Regulate Neuroendocrine Function

Bret N. Smith and F. Edward Dudek

Contents

0-8493-3363-6/98/$0.00+$.50
© 1998 by CRC Press LLC

I. Introduction

Neuroendocrine cells represent a class of central neurons that release peptides directly into the circulatory system. Release of these hormones depends on action potentials, and is therefore controlled by the electrical activity of the neuroendocrine cell. Both the pattern and frequency of action potential discharge are critical factors determining the amount of hormone released.[1,2] As with other central nervous system neurons, action potential generation in neuroendocrine cells results from the integration of intrinsic membrane properties and extrinsic inputs from neurotransmitters and modulators. The most direct method available to examine membrane properties in neurons is intracellular recording with either sharp electrodes or patch pipettes. Electrophysiological analysis of neuroendocrine cells is essential to understand the integrative mechanisms that ultimately control hormone release. This chapter will consider the utility and limitations of different methods for investigating electrical activity in neuroendocrine cells. Some of the important questions addressed by intracellular recordings will be examined, and we will discuss our views of how new techniques and combinations of methods will allow significant new insights into the cellular mechanisms of neuroendocrine function.

II. Recording Techniques

Early electrophysiological examination of the function of neuroendocrine neurons utilized extracellular recordings obtained from anesthetized animals, which yielded significant data concerning the types of stimuli effective in changing the firing behavior of neuroendocrine neurons. Most electrophysiological studies of neuroendocrine cells have been directed at the magnocellular system, which consists primarily of neurons in the supraoptic and paraventricular nuclei that secrete oxytocin and vasopressin into the neurohypophysis. They have indicated that action potential frequency, discharge pattern, and synchronization are critical for homeostatic and stimulated hormone release. Assessment of drug and other stimulus-induced effects on neuroendocrine cell function using extracellular recordings is, however, limited to changes in action potential frequency. Subthreshold events cannot be detected, and the pre- or postsynaptic actions of various drugs cannot be determined directly. Further, the contribution of intrinsic membrane properties to the firing behavior of neuroendocrine cells requires a more direct method of analysis.

Intracellular recordings provide information about the intrinsic properties of a neuron, make possible direct assessment of the actions of pharmacological agents (e.g., neurotransmitters and neuromodulators), and allow analysis of direct and indirect synaptic inputs. Intracellular electrophysiological recordings from magnocellular neuroendocrine cells (MNCs) have yielded valuable information pertaining directly to the mechanisms underlying neuroendocrine cell function. The magnocellular system has been the model for investigating neuroendocrine cell activity, but the principles guiding examination of parvocellular neuroendocrine cells controlling adenohypophysial hormone release are presumably similar to those for MNCs. Widely used intracellular recording techniques include single-electrode current- and voltage-clamp recordings with either sharp electrodes or patch pipettes. Each of these modes has distinct advantages and disadvantages, and each can be used to answer different kinds of questions.

A. Sharp-Electrode Current-Clamp Recording

Voltage variations that contribute to neuron excitability (e.g., action potentials, afterpotentials, and postsynaptic potentials) are most directly studied with so-called "current-clamp" recordings. An important issue in the study of neuroendocrine cell function has been to determine the contributions of intrinsic and extrinsic factors in generating action potential patterns.[3-7] Further, associating one or more of these factors with a particular neuron type has been a goal of several current-clamp recording studies.[8] Current-clamp recordings have been utilized in neuroendocrine cells to measure passive and active membrane properties, and the modulation of those properties by synaptic and pharmacological stimuli (Figure 1).

Current-clamp recordings have demonstrated the importance of excitatory (glutamate) and inhibitory (GABA) synaptic input to the function of neuroendocrine cells. For example, synaptic activation by glutamatergic neurons can render MNCs more excitable. This excitability can be a transient depolarization in the membrane potential due to binding of kainate/AMPA receptors,[9,10] resulting in action potentials. In addition, NMDA receptor activation depolarizes most neurons,[10-12] making them more likely to fire clusters of spikes.[13] Conversely, GABA usually inhibits MNCs by transiently hyperpolarizing them.[14]

Current-clamp recordings have also demonstrated the importance of intrinsic membrane properties in controlling action potential frequency and patterns. Passive membrane properties — including resting membrane potential, input resistance, cell capacitance, and electrotonic length of dendritic processes — contribute to the ability of a neuroendocrine cell to respond to synaptic input.[15] The effect of synaptic input on firing behavior probably depends as much on the pattern of that input as on it's soma-dendritic location. Active membrane properties, such as the depolarizing after-potential (Figure 1) and the afterhyperpolarization that follow spikes or trains of spikes in MNCs, can also render them more or less likely to fire additional action potentials.[3-5] The depolarizing afterpotential in particular has been associated with the ability of putative vasopressinergic neurons to fire phasic bursts of action potentials, although it is now clear that many oxytocin neurons also express a depolarizing afterpotential.[16] Although no single membrane property has been uniquely correlated to a specific neuron type in male rats *in vitro* (but see Reference 17), intrinsic membrane properties play critical roles in generating firing behaviors in neuroendocrine cells.

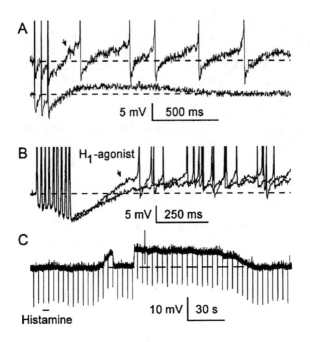

FIGURE 1

Depolarizing afterpotentials and the modulatory actions of histamine on membrane properties of MNCs in the supraoptic nucleus. (A.) Examples of a depolarizing afterpotential (arrow) following a series of three action potentials. Action potentials evoked from two membrane potentials (–55 mV, upper trace and –58 mV, lower trace) resulted in a depolarizing afterpotential in many MNCs. (B.) H_1-receptor activation enhanced the depolarizing afterpotential in identified vasopressin neurons. Averaged traces (n = 4) at identical membrane potentials (–62 mV; dashed line) prior to and after a bolus application of the H_1-agonist, 2-thiazolylethylamine (100 μM; arrow) are shown. (From Smith and Armstrong, *Neuroscience*, 53, 855, 1993. With permission.) (C.) Depolarization of an immunohistochemically identified vasopressinergic MNC by histamine. A bolus application of histamine depolarized this neuron ~7 mV from it's resting membrane potential of –64 mV, indicated by the dashed line. Hyperpolarizing current pulses were given to monitor input resistance. (From Smith, B.N. and Armstrong, W.E., *J. Physiol. (London)*, 495, 465, 1996. With permission.)

Finally, current-clamp recordings have been used extensively in MNCs to examine the effects of neuroactive substances, such as histamine (Figure 1), that alter neuronal excitability more slowly than GABA and glutamate (on the order of seconds). Such substances are therefore often referred to as neuromodulators. These neuromodulators are presumed to be contained in and released by terminals or axons near neuroendocrine cells. Their actions on neuroendocrine cells may therefore represent relevant physiological stimuli. Since neuromodulator effects can be indirect (i.e., non-synaptic or presynaptic) or subthreshold for action potential generation, they cannot always be accurately assessed by extracellular recordings. For example, amines and peptides can depolarize or hyperpolarize the membrane, depending on the receptor type activated.[8,18] Some neuroactive substances can also act postsynaptically to alter afterpotentials directly[19] and presynaptically to alter the amount of amino acid transmitter released at a synapse.[20] Although responses of different neurons to these agents have yet to provide a reliable way of differentiating between neuron types, their effects on membrane potential and synaptic input can be assessed by sharp-electrode current-clamp recordings.

Changes in membrane potential result in the activation and inactivation of voltage-dependent ion channels responsible for the action potential. The amplitude and duration of these voltage fluctuations are critically determined by the combination of ionic currents activated at a given time. One limitation of current-clamp recordings is the relative inability to control for activation of voltage-dependent conductances. Although current-clamp recordings can be used to study changes in membrane potential, they are ineffective in isolating voltage-dependent responses.

B. Voltage-Clamp Recording

Voltage-clamp recordings have been used in MNCs to examine activation, inactivation, and gating properties of voltage-dependent conductances underlying the potentials observed in current-clamp mode (Figure 2). They have also been used to study neuromodulator-induced changes in resting membrane current[21,22] and postsynaptic currents activated by amino acid neurotransmitters.[23] In voltage-clamp, the voltage of the neuron is theoretically held constant while measuring the transmembrane current necessary to maintain the clamped voltage. The advantage of voltage-clamp recording is that membrane conductances can be isolated from the capacitive current needed to charge the cell capacitance, which can be eliminated from the recording. Using this approach, ionic currents in MNCs have been identified that are common to many neurons throughout the brain.[24,25] The characteristics of other currents, such as those underlying the depolarizing afterpotential,[26] have been described but incompletely characterized.

One advantage of this technique over current-clamp recordings is that subtle effects of neuromodulators on specific currents can be isolated. This allows for a more mechanistic understanding cf how a given substance might alter neuronal behavior. A caveat to this technique is the ubiquitous "space-clamp" problem. This means that voltage-clamping the soma does not guarantee against the possibility that activity in unclamped dendrites influences apparent somatic current. Although neuroendocrine cells have less extensively arborized dendritic processes than many other neuron types, the contribution of the dendrites to activity seen at the soma is likely to be significant.[15,27] Nevertheless, voltage-clamp recordings have been used to identify several voltage- and neuromodulator-activated currents important for neuroendocrine cell function. This is especially true in neurons with minimal dendritic arborization, such as acutely dissociated[27,28] or cultured neurons.[29,30]

C. Patch-Clamp Recording

Both current- and voltage-clamp studies have been carried out in neuroendocrine cells using whole-cell patch-clamp recordings.[23,29-31] Patch pipettes have a relatively low tip resistance (2 to 6 MΩ is common) that allows for recordings with very low

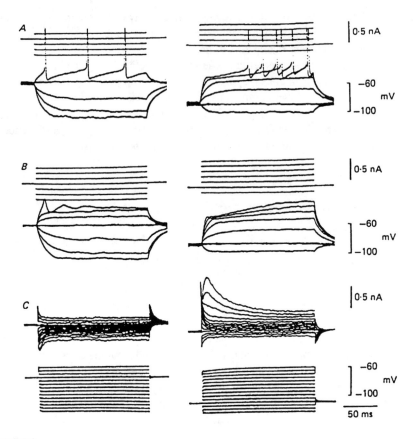

FIGURE 2

Voltage-clamp correlates of an intrinsic membrane property of MNCs. (A and B.) Current-clamp record-ings obtained from an MNC illustrate that the outwardly rectifying behavior, which appeared during depolarizing pulses applied from hyperpolarized potentials in control solutions (A) is not blocked by the addition of 0.5 μM tetrodotoxin (B). This treatment abolished Na^+-dependent action potentials. (C.) The membrane current (upper) responses of the same cell to voltage-clamp (lower) commands spanning a similar range of potentials. Note that a fast-activating transient outward current was elicited during the depolarizing commands applied from negative, but not positive, holding potentials. (From Bourque, *J. Physiol. (London)*, 397, 331, 1988. With permission.)

access resistance. Consequently, patch-clamp recordings provide good control of a larger portion of the membrane than offered by sharp-electrode recordings, reducing space-clamp problems. Small and rapid currents, such as quantal synaptic events (Figure 3), can be readily resolved with patch-clamp recordings.[23] Patch pipettes also allow ionic control of the intracellular milieu because the large tip diameter allows dialysis of the neuron with the pipette contents. This is also a distinct disadvantage because of the potential washout of intracellular constituents, espe-cially molecules required for second messenger-mediated activity and calcium-dependent events. These problems have been partially alleviated by adding adenosine triphosphate (ATP) or guanosine triphosphate (GTP) and adjusting the calcium

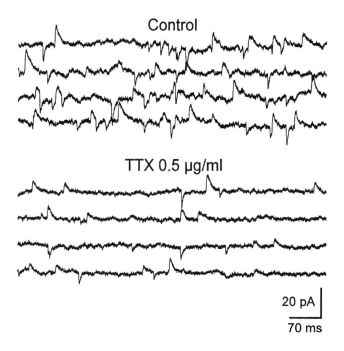

FIGURE 3

Whole-cell patch-clamp recordings from an MNC in the supraoptic nucleus comparing spontaneous synaptic activity in normal perfusion solution (top) and 20 min after the addition of tetrodotoxin (TTX; bottom) to block action potential-dependent synaptic activity. The cell input resistance was 500 MΩ. Traces are continuous, and holding potential was the same as resting membrane potential (–55 mV). (From Wuarin and Dudek, *J. Neurosci.*, 13, 2323, 1993. With permission.)

buffering capacity of the recording pipette solution.[31] Although patch-clamp recordings were initially done on dissociated neurons and cell cultures, the application of the technique to slices has been successful in many brain areas, including the hypothalamus.

In slices, where some local synaptic circuitry is intact, the high resolution of the patch-clamp technique allows quantitative analysis of miniature postsynaptic currents (i.e., due to action potential-independent release of neurotransmitter; Figure 3). Using patch-clamp recordings in slices, the contribution of local circuit neurons to the synaptic profile of neuroendocrine cells can be determined. This type of recording has established that local amino acid-mediated synaptic transmission is ongoing in neuroendocrine cells.[23] The frequency and patterning of excitatory or inhibitory inputs are presumably important for determining the activity of neuroendocrine cells. The anatomical location of many of these inputs is yet to be determined, and the pre- and postsynaptic effects of neuromodulators on spontaneous amino acid-mediated events are also largely unknown. Whole-cell patch-clamp recordings will be especially useful for determining how synaptic activity regulates neuroendocrine function.

III. Preparations

Although a few intracellular recordings from neuroendocrine neurons have been done in intact animals,[32,33] the majority have been performed *in vitro*. *In vitro* preparations have several advantages over intact animals for intracellular and whole-cell recordings. In addition to greater stability, *in vitro* preparations allow visualization of the region to be recorded and control of the extracellular environment (e.g., ion and drug concentrations). The most commonly used preparations have been "thick" slices (~500 μm) and explants. Recordings have also been obtained in acutely dissociated and cultured MNCs. Each of these preparations has distinct advantages and disadvantages, as discussed below. The recent development of the "thin-slice" (~150 to 250 μm) preparation offers another option for recording from neuroendocrine neurons that combines several of the advantages of these earlier-developed preparations.

A. The "Thick Slice" Preparation

Traditional "thick slice" preparations have been used with all of the recording techniques discussed above (Figure 4). These preparations are especially amenable to electrical stimulation of input pathways and identification of drug effects on MNCs. Explants, essentially very thick (~800 to 1200 μm) horizontal slices of the ventral hypothalamus (Figure 4), have also been used to examine rostro-caudal intrahypothalamic connections and to identify MNCs in the supraoptic nucleus by antidromically stimulating the neural lobe or stalk.[7,34,35] A principal advantage of traditional slices is that neurons are often largely intact, and many local synaptic connections are preserved. This makes the thick slice a good preparation for examining the effects of local synaptic input and also for comparing anatomical features with intrinsic membrane properties of relatively intact neuroendocrine cells.

B. Isolated Neurons

The electrophysiological characteristics of MNCs have been examined in both cultured and dissociated preparations. Hypothalamic neuron cultures have been used to examine developmentally regulated membrane properties.[29,30,36] In addition, synapses can form between cultured neurons, making possible visually-guided recordings during periods of synaptogenesis.[20] Acutely dissociated MNCs have been used for intracellular investigation of intrinsic membrane properties[27,28] and postsynaptic neurotransmitter effects.[27] The techniques applied to these preparations include whole-cell patch-clamp, single-channel recordings in isolated patches, calcium imaging, and intracellular recording. Dissociated neurons usually have little if any dendritic arborization, which minimizes space-clamp considerations. This makes them

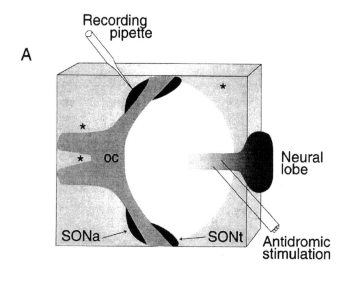

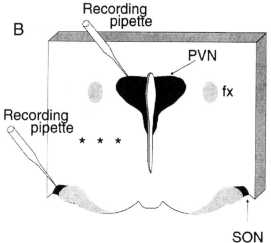

FIGURE 4

Schematic diagrams of *in vitro* hypothalamic slice preparations. (A.) Ventral surface of a hypothalamo-neurohypophysial explant. Both the anterior supraoptic nucleus (SONa) and the tuberal supraoptic nucleus (SONt) are accessible. An intracellular recording electrode is positioned in the SONa. The pituitary stalk can be stimulated to trigger antidromic action potentials, and the areas indicated by asterisks (*) can be stimulated to evoke synaptic responses. The optic nerves, tracts, and chiasm (oc) serve as landmarks. (B.) Coronal hypothalamic slice. Intracellular recordings can be obtained from the supraoptic nucleus (SON), paraventricular nucleus (PVN), or other hypothalamic nuclei with the slice preparation. Ortho-dromic stimulation can be applied at several sites near the recording electrode, as indicated by the asterisks (*). The optic tracts, third ventricle, and fornix (fx) serve as landmarks for positioning electrodes.

useful for assessing intrinsic currents, but electrophysiological results obtained in this preparation are sometimes different from those observed in slices or cultured cells. For example, close to 50% of MNCs in the supraoptic nucleus exhibit a depolarizing afterpotential in slices,[16,19,37] and most cells express an A-type current in explants[16,24] and in cultures.[29] However, relatively few dissociated MNCs have a depolarizing afterpotential,[27] and only oxytocin cells express the A-type current in dissociated MNCs.[38] One hypothesis for these differences is that channels located on the dendrites of MNCs contribute to the generation of these events. Simultaneous somatic and dendritic recordings in MNCs, as have been performed in neocortical neurons,[39] might be used to address this hypothesis. Thus, although intrinsic membrane properties of MNCs can vary between preparations, the differences in themselves suggest important electrophysiological properties of neuroendocrine cells.

C. The "Thin Slice" Preparation

Another type of slice preparation, the "thin slice", allows neurons to be visualized at high magnification under a compound microscope for whole-cell and isolated patch recordings.[40] Thin slices have typically been made from young (<4 wk old) animals. The utility of this preparation for recording from neuroendocrine cells has not yet been fully investigated. However, as discussed below, the benefit of *a priori* identification of neuroendocrine cells in slices could prove to be great. Thin slices allow many of the techniques used in cultured and dissociated neurons to be applied to slices, where at least some local circuitry remains intact and many of the morphological features of neurons are preserved. For example, single-channel analysis of isolated membrane patches can be performed in thin slices,[41] as can investigation of local synaptic integration. One disadvantage of slice preparations is that some neurons at the surfaces of the slice are damaged during preparation. This can be problematic in thin slices, since the visually guided recordings are usually restricted to regions near the slice surface. While it is likely that better visualization methods, such as infra-red microscopy,[39] will allow neuron visualization deeper into slices and in thicker slices, these methods have not yet been applied to neuroendocrine cells. Thick and thin slices are therefore useful for analysis of intrinsic membrane properties and of local synaptic control of neuroendocrine function, whereas isolated neurons are better suited to examine intrinsic currents or developmentally regulated properties of neuroendocrine cells.

IV. Stimulation Techniques

In slices, extracellular stimulation can be used to activate synaptic inputs or antidromically activate neuroendocrine cells. These techniques are best applied to identify the neuroendocrine nature of the recorded neuron or to the study of local synaptic circuitry, an important component of the regulation of neuroendocrine cell activity.

A. Electrical Stimulation

In slices, electrical stimulation of a variety of regions activates glutamatergic and/or GABAergic input[10,42] to the paraventricular or supraoptic nuclei. These types of studies have provided important information concerning the responses of MNCs to synaptically-released neurotransmitters. Electrical stimulation in brain slices activates intact neurons and also fibers from distant neurons; the origin of these fibers is usually unknown. It is therefore well suited for investigating postsynaptic responses and the effects of drugs on those responses, but not necessarily for identification of specific neuronal inputs.

B. Glutamate Stimulation: Glutamate Microdrops

One method of orthodromic stimulation that is likely to activate selectively neuronal somata and dendrites in the slice is local application of the excitatory amino acid, glutamate.[43,44] When applied to slices as microdrops, glutamate stimulates intact neurons, causing them to fire action potentials. Such stimulation results in activation of neurons (but not axons of passage) that project, either directly or via an interneuron, to the recorded neuron (Figure 5). Tasker and co-workers[43,44] used glutamate stimulation to map the origins of local GABAergic input to neuroendocrine cells in the paraventricular nucleus (Figure 5). Because the temporal and spatial resolution of the glutamate microdrop technique is low due to the diffusion of the glutamate through and across the surface of the slice, it is difficult to assess with precision the individual inputs to a recorded neuron. The identity of local circuitry mediating neuroendocrine activity has therefore begun to be defined, but is largely incomplete.

C. Glutamate Stimulation: Photolysis of Caged Glutamate

Glutamate photostimulation is a powerful tool for investigating local connectivity, both because of the relative specificity associated with somatic activation and because of the good spatial and temporal resolution of glutamate photolysis. This technique uses a flash of ultraviolet light (xenon flash lamp or laser) to release photolabile glutamate molecules (γ-(a-carboxy-2-nitrobenzyl) ester, trifluoroacetate (γ-(CNB-caged)) into relatively discrete regions of the slice.[45] Using a flash lamp, the spatial resolution of the glutamate photoactivation can be focused to relatively small areas (100 to 200 μm in diameter) by adjusting both the concentration of caged glutamate and the amount of flash energy used for uncaging. The spatial resolution can be increased by at least an order of magnitude if a laser is used to focus the light. In addition, the temporal resolution of this technique is very good, with the time between the flash of UV light and resulting postsynaptic activity on the order of milliseconds, vs. up to seconds with the glutamate microdrop. This method should

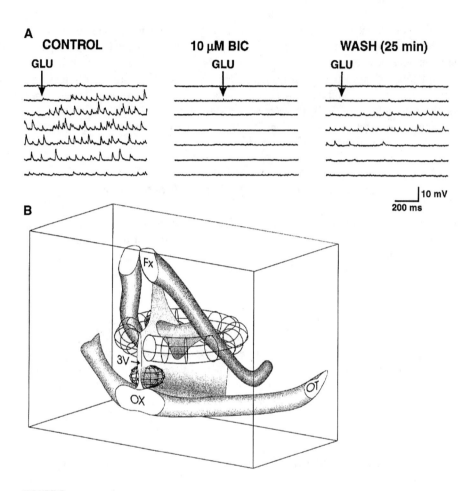

FIGURE 5

Glutamate microstimulation of inhibitory input to paraventricular nucleus neurons. (A.) *Left*, Glutamate microstimulation (GLU) at a position ventral to the fornix elicited reversed IPSPs in a paraventricular parvocellular neuron recorded with a KCl-filled microelectrode. *Middle*, Bath application of the GABA$_A$-receptor antagonist bicuculline methiodide (BIC) for 15 min completely blocked the effect of glutamate microapplication at the same site and with the same application parameters. *Right*, Partial recovery of the glutamate-evoked IPSPs was seen after washout of the bicuculline. The membrane potential was held at –105 mV with negative current injection. (B.) A qualitative schematic diagram depicting an approximated three-dimensional topography of zones concentrating paraventricular nucleus-projecting GABA neurons. Fx, fornix; OT, optic tract; OX, optic chiasm; PVN, paraventricular nucleus; 3V, third ventricle. (From Boudaba et al., *J. Neurosci.*, 16, 7151, 1996. With permission.)

prove useful for identifying discrete groups of locally projecting neurons. One principal issue that can be addressed using glutamate photostimulation is the degree to which individual types of neuroendocrine cells are differentially or commonly regulated by local synaptic input.

V. Neuron Identification

Of paramount importance for correlating results from intracellular recordings with possible physiological relevance (i.e., release of hormone) is identification of the peptide content of the recorded neuroendocrine cell. Two questions usually need to be answered: (1) Is it a neuroendocrine cell? and (2) What type of neuroendocrine cell is it?

A. Is it Neuroendocrine?

1. *Post-hoc* Anatomical Verification

Injecting neurons with a dye or other intracellular label can be used to identify the location and morphology of recorded neuroendocrine cells. Recently, biocytin[46] or neurobiotin[47] have been the labels of choice for identifying recorded MNCs, although Lucifer Yellow[48,49] and ethidium bromide can also be employed successfully.[50] Biotin-filled neurons can be histochemically identified by reacting the tissue with avidin-containing fluorophores or horseradish peroxidase conjugates for subsequent visualization with a chromogen, such as diaminobenzidine. Neurons that can be antidromically activated from the neural stalk or neurohypophysis are most likely MNCs, especially if their axons can be identified anatomically in these same areas.[35] Although marking filled neurons allows identification of their location and morphology, no anatomical features are currently known to be correlated with the identity of specific peptidergic neuron types.

2. *A priori* Anatomical Identification

Since most of the MNCs in the supraoptic and paraventricular nuclei are either vasopressin- or oxytocin-producing cells, electrodes placed in these areas will probably result in recordings from neuroendocrine cells. However, *a priori* identification of other neuroendocrine cells in slices has not been possible. Several years ago, Weiss and Cobbett[51] investigated the efficacy, stability, toxicity, and cost-efficiency of several fluorescent tracers injected intravenously in order to retrogradely label neuroendocrine cells via the neuro-humeral junction at their terminals. Neuroendocrine cells were then identified and recorded after acute dissociation, suggesting the possibility of applying this technique to label neuroendocrine cells in thin slices. Since many non-neuroendocrine cells reside in the same regions as neuroendocrine cells, especially in the tuberoinfundibular system, identification of neuroendocrine cells prior to recording would increase greatly the efficiency of data collection from identified neuroendocrine neurons.

B. What Type of Neuroendocrine Cell is it?

Unequivocal identification of neuroendocrine cells based on electrophysiological properties in slices has been largely unsuccessful.[8] The physiological state and gender of the animal can affect membrane properties,[17] further complicating the issue of correlating electrophysiological properties with peptidergic content. One means of neuron identification has been to combine intracellular labeling with immunohistochemical identification of the peptidergic content of MNCs[16,19,34,49,52-54] and parvocellular neurons.[55] In MNCs, it has been shown to be important to identify the specific neurophysins for both oxytocin and vasopressin in any given neuron (Figure 6), because some neurons are immunoreactive for neither peptide, and some are immunopositive for both.[16,19,53] Further, since recordings are usually obtained from more than one type of neuron, positive identification of peptide content increases the number of recordings that can be assigned to a specific neuron phenotype. Recordings from parvocellular neuroendocrine neurons of the tuberoinfundibular system can also benefit from this approach. A substantially larger number of peptide types are found in this system, and the different neuron types are often not anatomically segregated from each other or from non-neuroendocrine cells, as are the MNCs. Identification of parvocellular neuroendocrine cells is therefore a more complex issue than it is for MNCs.

In the parvocellular system, where several candidate peptides can exist in close proximity, it may be necessary to apply several antibodies to a single intracellularly labeled neuron. One technique that has been applied to MNCs is to embed the section containing the recorded and filled neuron in plastic, and then cut several semi-thin (1 to 2 μm) sections through the soma of the neuron of interest. Adjacent sections containing the filled neuron can be placed on separate slides, and can then be individually immunoreacted for one of several different peptides, as has been done for MNCs.[16,19,53] Other options for identifying recorded neuroendocrine neurons may eventually include using molecular biological techniques (e.g., *in situ* hybridization or polymerase chain reaction) to determine the phenotype of the recorded neuron. Ultimately, identification of the peptide content of recorded neuroendocrine cells is necessary to correlate electrophysiological data with the neuroendocrine function of a given neuron.

VI. Future Directions

Among the important issues yet to be resolved with recordings from neuroendocrine cells is the relative contribution of extrinsic and intrinsic mechanisms in controlling their firing behavior. Although fundamental membrane properties are fairly well established for MNCs, very little is known about the intrinsic properties of any single parvocellular neuroendocrine cell type.[55] Similarly, control of neuroendocrine function by local circuit interactions has only begun to be examined.[43,44] Identifying mechanisms that are selective or common to the various types of neuroendocrine cells will also be important in the context of understanding the release of any particular hormone. Since such release (and presumably neuron activity) is gender- and state-dependent,[17] more than one electrophysiological profile may be required

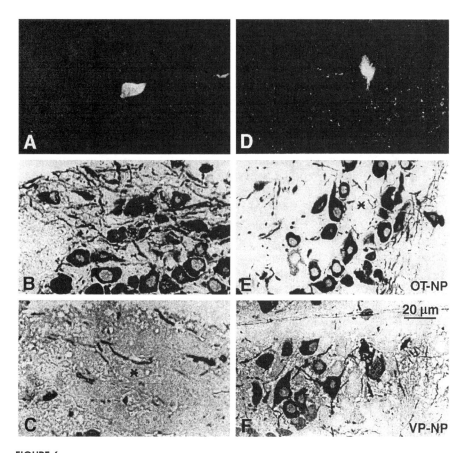

FIGURE 6

Examples of immunocytochemical localization of oxytocin- and vasopressin-neurophysin (OT-NP or VP-NP, respectively) in intracellularly labeled supraoptic nucleus neurons in adjacent 2 μm plastic sections. (A-C.) The filled neuron in A was identified as positive for OT-NP (B), and negative for VP-NP (C). (D-F.) Filled neuron in D is negative for OT-NP (E) and positive for VP-NP (F). In both cases, the neurons were filled with neurobiotin and reacted using avidin-Texas Red. The asterisk identifies the filled cell in the immunocytochemically stained sections. (From Armstrong et al., *J. Physiol. (London)*, 475, 115, 1994. With permission.)

for a given neuron type. Understanding fully the mechanisms underlying hormone release will probably require electrophysiological analysis of multiple aspects of neuronal control (e.g., intrinsic properties, local circuitry, and neuromodulator effects) under a variety of physiologically relevant conditions, depending on the neuroendocrine function being examined. This is especially true for the electrophysiology of neurons controlling adenohypophysial hormone release, about which very little is currently known. Development of a model system for studying parvocellular neuroendocrine function will greatly advance this effort.

VII. Conclusions

Intracellular recordings have been used to elucidate cellular mechanisms underlying the firing characteristics of neuroendocrine cells. Although the study of MNC electrophysiology has dominated the field, analogous techniques can be applied to study the behavior of other neuroendocrine cells. Major issues, such as the relative contributions of extrinsic and intrinsic mechanisms to neuronal behavior, can be addressed with intracellular recordings. The mechanisms underlying behavior of specific neuron types can be determined by combining intracellular recordings with anatomical and immunohistochemical identification of the recorded neuron. Since most recordings are performed using *in vitro* preparations, intracellular recordings are not currently applicable to study directly the effects of activity in many of the projection pathways that probably contribute to control of neuroendocrine function. However, the postsynaptic effects of neuroactive substances can be determined, and therefore indirect assessment of how major projections might influence a particular neuron can be made. Attempts to date to ascribe an electrophysiological "signature" to a given type of neuroendocrine cell *in vitro* have been only moderately successful, and this certainly remains one of the primary questions in the study of how their function is controlled. Further, the influences controlling function of parvocellular neurons of the tuberoinfundibular system are only beginning to be elucidated. New technology, combined with classical recording and staining techniques, may provide the necessary means to investigate the cellular mechanisms controlling neuroendocrine function.

Acknowledgments

We wish to thank Dr. W. E. Armstrong for providing the plate for Figure 6. This work was supported by a fellowship from the American Heart Association of Colorado, Inc. (BNS) and a grant from the AFSOR (FED).

References

1. Bicknell, R. J. and Leng, G., Relative efficacy of neuronal firing patterns for vasopressin release *in vitro*, *Neuroendocrinology*, 33, 295, 1981.
2. Cazalis, M., Dayanithi, G., and Nordmann, J. J., The role of patterned burst and intrerburst interval on the excitation-coupling mechanism in the isolated rat neural lobe, *J. Physiol. (London)*, 369, 45, 1985.
3. Andrew, R. D. and Dudek, F. E., Burst discharge in mammalian neuroendocrine cells involves an intrinsic regenerative mechanism, *Science*, 221, 1050, 1983.
4. Andrew, R. D. and Dudek, F. E., Analysis of intracellularly recorded phasic bursting by mammalian neuroendocrine cells, *J. Neurophysiol.*, 51, 552, 1984.

5. Andrew, R. D., Endogenous bursting by rat supraoptic neuroendocrine cells is calcium dependent, *J. Physiol. (London)*, 384, 451, 1987a.

6. Andrew, R. D., Isoperiodic bursting by magnocellular neuroendocrine cells in the rat hypothalamic slice, *J. Physiol. (London)*, 384, 467, 1987b.

7. Dudek, F. E., Tasker, J. G., and Wuarin, J.-P., Intrinsic and synaptic mechanisms of hypothalamic neurons studied with slice and explant preparations, *J. Neurosci. Meth.*, 28, 59, 1989.

8. Armstrong, W. E., Morphological and electrophysiological classification of hypothalamic supraoptic neurons, *Prog. Neurobiol.*, 47,291, 1995.

9. Gribkoff, V. K. and Dudek, F. E., Effects of the excitatory amino acid antagonists on synaptic responses of supraoptic neurons in slices of rat hypothalamus, *J. Neurophysiol.*, 63, 60, 1990.

10. Yang, C. R., Senatorov, V. V., and Renaud, L. P., Organum vasculosum lamina terminalis-evoked postsynaptic responses in rat supraoptic neurones *in vitro*, *J. Physiol. (London)*, 477, 59, 1994.

11. Dudek, F. E. and Gribkoff, V. K., Synaptic activation of a slow depolarization in rat supraoptic nucleus neurones *in vitro*, *J. Physiol. (London)*, 387, 273, 1987.

12. Gribkoff, V. K., Electrophysiological evidence for N-methyl-D-aspartate excitatory amino acid receptors in the rat supraoptic nucleus *in vitro*, *Neurosci. Lett.*, 131, 260, 1991.

13. Hu, B. and Bourque, C. W., NMDA receptor-mediated rhythmic bursting activity in rat supraoptic nucleus neurones *in vitro*, *J. Physiol. (London)*, 458, 667, 1992.

14. Randle, J. C. W. and Renaud, L. P., Actions of gamma-aminobutyric acid on rat supraoptic nucleus neurosecretory neurons *in vitro*, *J. Physiol. (London)*, 387, 629, 1987.

15. Armstrong, W. E. and Smith, B. N., Tuberal supraoptic neurons — II: Electrotonic properties, *Neuroscience*, 38, 485, 1990.

16. Armstrong, W. E., Smith, B. N., and Tian, M., Electrophysiological characteristics of immunochemically identified oxytocin and vasopressin neurons *in vitro*, *J. Physiol. (London)*, 475, 115, 1994.

17. Stern, J. E. and Armstrong, W. E., Electrophysiological differences between oxytocin and vasopressin neurones recorded from female rats *in vitro*, *J. Physiol. (London)*, 488, 701, 1995.

18. Renaud L. P. and Bourque, C. W., Neurophysiology and neuropharmacology of hypothalamic magnocellular neurons secreting vasopressin and oxytocin, *Prog. Brain Res.*, 36, 131, 1991.

19. Smith, B. N. and Armstrong, W. E., Histamine enhances the depolarizing afterpotential of immunohistochemically identified vasopressin neurons in the rat supraoptic nucleus via H_1-receptor activation, *Neuroscience*, 53, 855, 1993.

20. Obrietan, K., Belousov, A. B., Heller, H. C., and van den Pol, A. N., Adenosine pre- and postsynaptic modulation of glutamate-dependent calcium activity in hypothalamic neurons, *J. Neurophysiol.*, 74, 2150, 1995.

21. Jarvis, C. R., Bourque, C. W., and Renaud, L. P., Depolarizing action of cholesystokinin on rat supraoptic neurones *in vitro*, *J. Physiol. (London)*, 458, 621, 1992.

22. Yang, C. R., Bourque, C. W., and Renaud, L. P., Dopamine D_2 receptor activation depolarizes rat supraoptic neurones in hypothalamic explants, *J. Physiol. (London)*, 443, 405, 1991.

23. Wuarin, J.-P. and Dudek, F. E., Patch-clamp analysis of spontaneous synaptic currents in supraoptic neuroendocrine cells of the rat hypothalamus, *J. Neurosci.*, 13, 2323, 1993.

24. Bourque, C. W., Transient calcium-dependent potassium current in magnocellular neurosecretory cells of the rat supraoptic nucleus, *J. Physiol. (London)*, 397, 331, 1988.

25. Erickson, K. R., Ronnekleiv, O. K., and Kelly, M. J., Inward rectification (I_h) in immunocytochemically-identified vasopressin and oxytocin neurons of guinea-pig supraoptic nucleus, *J. Neuroendocrinol.*, 2, 261, 1990.

26. Bourque, C. W., Calcium-dependent spike after-current induces burst firing in magnocellular neurosecretory cells, *Neurosci. Lett.*, 70, 204, 1986.

27. Oliet, S. H. R. and Bourque, C. W., Properties of supraoptic magnocellular neurones isolated from the adult rat, *J. Physiol. (London)*, 455, 291, 1992.

28. Cobbett, P. and Weiss, M. L., Voltage-clamp recordings from identified dissociated neuroendocrine cells of the adult rat supraoptic nucleus, *J. Neuroendocrinol.*, 2, 267, 1990.

29. Cobbett, P., Legendre, P., and Mason, W. T., Characterization of three types of potassium current in cultured neurones of rat supraoptic nucleus area, *J. Physiol. (London)*, 410, 443, 1989.

30. Muller, T. H., Misgeld, U., and Swandulla, D., Ionic currents in cultured rat hypothalamic neurones, *J. Physiol. (London)*, 450, 341, 1992.

31. Li, Z. and Hatton, G. I., Histamine-induced prolonged depolarization in rat supraoptic neurons: G-protein-mediated, Ca^{2+}-independent suppression of K^+ leakage conductance, *Neuroscience*, 70, 145, 1996.

32. Bourque C. W. and Renaud, L. P., Membrane properties of rat magnocellular neuroendocrine cells *in vivo*, *Brain Res.*, 540, 349, 1991.

33. Dyball, R. E. J., Tasker, J. G., Wuarin, J.-P., and Dudek, F. E., *In vivo* intracellular recording of neurons in the supraoptic nucleus of the rat hypothalamus, *J. Neuroendocrinol.*, 3, 383, 1991.

34. Reaves, T. A., Hou-Yu, A., Zimmerman, E. A., and Hayward, J. N., Supraoptic neurons in the rat hypothalamo-neurohypophysial explant: Double-labeling with Lucifer Yellow injection and immunocytochemical identification of vasopressin- and neurophysin-containing neuroendocrine cells, *Neurosci. Lett.*, 37, 137, 1983.

35. Smith, B. N. and Armstrong, W. E., Tuberal supraoptic neurons — I: Morphological and electrophysiological characteristics observed with intracellular recording and biocytin filling *in vitro*, *Neuroscience*, 38, 469, 1990.

36. van den Pol, A. N., Obrietan, K. Cao, V., and Trombley, P. Q., Embryonic hypothalamic expression of functional glutamate receptors, *Neuroscience*, 67, 419, 1995.

37. Li, Z. and Hatton, G. I., Ca^{2+} release from internal stores: Role in generating depolarizing afterpotentials in rat supraoptic neurones, *J. Physiol. (London)*, 498, 339, 1997.

38. Widmer, H., Boissin-Agasse, L., Richard, P., and Desarmenien, M. G., Differential distribution of a potassium current in immunocytochemically identified supraoptic magnocellular neurones of the rat, *Neuroendocrinology*, 65, 229, 1997.

39. Stuart, G. J. and Sakmann, B., Active propagation of somatic action potentials into neocortical pyramidal cell dendrites, *Nature (London)*, 367, 69, 1994.

40. Edwards, F. A., Konnerth, A., Sakmann, B., and Takahashi, T., A thin slice preparation for patch clamp measurements on identified neurones of the mammalian central nervous system, *Pflugers Arch.*, 414, 600, 1989.

41. Strecker, G. J., Jackson, M. B., and Dudek, F. E., Blockade of NMDA-activated channels by magnesium in the immature rat hippocampus, *J. Neurophysiol.*, 72, 1538, 1994.

42. Gribkoff, V. K. and Dudek, F. E., The effects of the excitatory amino acid antagonist kynurenic acid on synaptic transmission to supraoptic neuroendocrine cells, *Brain Res.*, 442, 152, 1988.

43. Boudaba, C., Szabo, K., and Tasker, J. G., Physiological mapping of local inhibitory inputs to the hypothalamic paraventricular nucleus, *J. Neurosci.*, 16, 7151, 1996.

44. Tasker, J. G. and Dudek, F. E., Local inhibitory synaptic inputs to neurones of the paraventricular nucleus in slices of rat hypothalamus, *J. Physiol. (London)*, 469, 179, 1993.

45. Callaway, E. M. and Katz, L. C., Photostimulation using caged glutamate reveals functional circuitry in living brain slices, *Proc. Natl. Acad. Sci. USA,* 90, 7661, 1993.

46. Horikawa, K. and Armstrong, W. E., A versatile means of intracellular labeling: injection of biocytin and its detection with avidin conjugates, *J. Neurosci. Meth.*, 25, 1, 1988.

47. Kita, H. and Armstrong, W. E., A biotin-containing compound N-(2-aminoethyl) biotinamide for intracellular labeling and neuronal tracing studies: comparison with biocytin, *J. Neurosci. Meth.*, 37, 141, 1991.

48. Tasker, J. G. and Dudek, F. E., Electrophysiological properties of neurones in the region of the paraventricular nucleus in slices of rat hypothalamus, *J. Physiol. (London)*, 434, 271, 1991.

49. Yamashita, H., Inenaga, K., Kawata, M., and Sano, Y., Phasically firing neurons in the supraoptic nucleus of the rat hypothalamus: Immunocytochemical and electrophysiological studies, *Neuroscience*, 37, 87, 1983.

50. Tasker, J. G., Hoffman, N. W., and Dudek, F. E., A comparison of three intracellular markers for combined electrophysiological, anatomical and immunohistochemical analyses, *J. Neurosci. Meth.*, 38, 129, 1991.

51. Weiss, M. L. and Cobbett, P., Intravenous injection of evans blue labels magnocellular neuroendocrine cells of the rat supraoptic nucleus *in situ* and after dissociation, *Neuroscience*, 48, 383, 1992.

52. Cobbett, P., Smithson, K. G., and Hatton, G. I., Immunoreactivity to vasopressin- and not oxytocin-associated neurophysin antiserum in phasic neurons of rat hypothalamic paraventricular nucleus, *Brain Res.*, 362, 7, 1986.

53. Smith, B. N. and Armstrong, W. E., The ionic dependence of the histamine-induced depolarization of vasopressin neurones in the rat supraoptic nucleus, *J. Physiol. (London)*, 495, 465, 1996.

54. Yang, Q. Z. and Hatton, G. I., Histamine mediates fast synaptic inhibition of rat supraoptic oxytocin neurons via chloride conductance activation, *Neuroscience*, 61, 955, 1994.

55. Lagrange, A. H., Ronnekleiv, O. K., and Kelly, M. J., Estradiol-17β and μ-opioid peptides rapidly hyperpolarize GnRH neurons: A cellular mechanism of negative feedback?, *Endocrinology*, 136, 2341, 1995.

Chapter

Application of *In Vivo* Electrophysiological Approaches to the Study of Neuroendocrine Systems

Joan M. Lakoski and Jane E. Smith

Contents

I. Introduction

Electrophysiological recording techniques have been recognized to provide a "window" on the cellular dynamics of peripheral and central nervous system tissues. As applied to investigations of neuroendocrine system function, these approaches have facilitated the assessment of receptor/effector coupling processes that are intrinsic to hormone secretion and feedback responses on nervous system tissues. Whether under conditions of a controlled endocrine environment, such as achieved with hormone replacement therapy, or in the intact animal responding to environmental or exogenous drug-induced stimuli, the cellular physiological responses that mediate the cascade of events regulating a neuroendocrine system can be elucidated with the application of electrophysiological approaches. While the isolated neuroendocrine cell and the *in vitro* brain slice preparation are well suited for intracellular recording and other membrane biophysical techniques that require a high degree of mechanical stability of the recording preparation (see Chapter 6), the *in vivo* approach offers the advantage of intact nervous and endocrine systems that can interact concordantly. Likewise, the systems approach inherent in the use of *in vivo* recording techniques provides information on the circuitry of a given neuroendocrine system, including identification of responses in selected brain regions and cell populations.

This chapter addresses the uses and limitations of cellular electrophysiological recording methods for investigating the interaction of nervous and endocrine systems. We will survey a variety of electrophysiological approaches useful in the intact animal for determination of hormone-, neurotransmitter- and modulator-mediated responses which underlie the processes of cellular integration in neuroendocrine systems.

II. Recording Preparations

A benefit of critical importance to the determination of physiological and pharmacological responses in a neuroendocrine system is that the organism and hence, the neuroendocrine system, is intact. Therefore, using an *in vivo* recording preparation it is possible to examine the function of an individual neuron within the context of a natural fluctuating hormonal environment. Moreover, the anatomical circuitry of the cell under investigation as well as the peripheral target organs for a given endocrine system are available to respond and adapt to changes in the hormone milieu. As such, use of an *in vivo* preparation for electrophysiological experiments provides several distinct advantages over *in vitro* brain slice or isolated cell culture preparations. A clear limitation to *in vitro* systems for physiological studies is the difficulty in ascertaining that the preparation faithfully reflects the biology of a living organism. Often the factors, modulators or intracellular components vital to the receptor/effector coupling process under investigation simply diffuse from the cell of interest into the recording chamber.

As applied to neuroendocrine investigations, the *in vivo* recording preparation is a feasible model for extracellular recording studies of single or multiple cells. This approach can be used to identify the neuroanatomical circuitry that regulates the responsiveness of a given hormone-dependent response. Extracellular recordings can be conducted in animals in which a discrete knife cut has been placed to eliminate an anatomical input or output. Alternatively, recordings can be conducted following placement of a neurotoxin in the brain in order to produce selective neuronal damage, chronic drug treatment and/or hormone exposure.[2,3] Likewise, endocrinologically relevant hormone agonists or antagonists can be infused into the lateral ventricles (intraventricular injection) or relevant brain site (intracerebral injection) while recording from the target of interest. The *in vivo* preparation readily allows for comparisons of a given tissue or brain region under multiple endocrine conditions which are relevant to questions regarding the consequences of prolonged hormone exposure on receptor-mediated responses.

It is important to acknowledge that the use of an *in vivo* preparation requires that careful and humane consideration be given to the animal used in biomedical research. Approval for the use of animals in a research protocol must be obtained from the investigator's Institutional Animal Care and Use Committee and should comply with federal guidelines.

A. Anesthetized Recording Preparations

The choice of anesthetic agent is an important aspect of the experimental design and should be based on several considerations. First, the duration of the anesthetic state needed to complete any surgical preparation and the recording session should be determined. Often, a more rapid surgical procedure which can be performed in minutes, such as an ovariectomy, can be accomplished using a volatile anesthetic agent such as halothane. However, for a recording session lasting several hours, the agents of choice are administered by intravenous or intraperitoneal routes; these routes of administration allow for easy supplementation at frequent intervals. Systemically administered agents that are commonly used as anesthetics in neuroendocrine studies of mammalian species include urethane, pentobarbital and chloral hydrate. For extracellular single-unit recording studies in the hippocampus and dorsal raphe nucleus in the female Fischer 344 rat, our laboratory routinely uses an initial dose of 400 mg/kg chloral hydrate (16% solution; intraperitoneal injection);[3,4,5] supplementation of the anesthetic is provided at regular intervals throughout the recording session via a cannulated lateral tail vein.

In choosing an anesthetic agent, it is important to recognize that the compound chosen may have effects on the firing characteristics and neuronal responsiveness of the population of neurons under investigation. For example, the patterns of spontaneous cell firing recorded in the locus coeruleus differ markedly in the anesthetized vs. freely moving recording preparation.[6,7] The sensitivity of the noradrenergic neurons in the locus coeruleus to application corticotropin-releasing hormone is also decreased in the presence of an anesthetic agent.[6] Other variables that are important to consider in choosing an anesthetic may include the gender and age of the subjects. For example, pentobarbital anesthetic-dependent differences have been reported in young vs. middle-aged female rats[8] while chloral hydrate anesthetic produces no significant changes in spontaneous cell firing rates recorded in the hippocampus and dorsal raphe nucleus in the same animal model.[3,5] Regardless of the cell population to be examined, it is important to ascertain that the anesthetic chosen does not interact with the endocrine system under investigation.

B. Freely Moving Recording Preparations

The freely moving recording preparation has provided key insights into the regulation of hormone secretion by hypothalamic neurons. While the animal is anesthetized for implantation of stimulating electrodes or guide cannula, the subject is unanesthetized and may be in multiple states of arousal during the recording session.[9,10,11] For example, the identification of pulsatile oxytocin-secreting cells in the magnocellular neurosecretory region of the paraventricular region was conducted in the lactating female rat, with the stimulus for hormone secretion provided by suckling pups.[12,13] These electrophysiological studies demonstrated that activation of this neuroendocrine system included the production of synchronous bursting of high frequency action potentials in this hypothalamic region immediately prior to the release of oxytocin from the pituitary. Extracellular multi-unit recordings in the mediobasal hypothalamus and arcuate nucleus of the rhesus monkey have demonstrated the vital role of neuronal cell firing in the regulation of gonadotropin-releasing hormone secretion.[14,15,16] It is also feasible to investigate neuronal activity during naturally occurring sleep-wake states with this approach[10] and this approach offers the advantage of being readily combined with microdialysis techniques to investigate the pharmacology of a given brain region.[9,11] This is an invaluable approach to garner information on an identified cell population over prolonged periods of time and typically requires the use of metal electrodes as described below.

III. Recording Techniques

A variety of recording techniques are currently available for *in vivo* studies of the electrophysiological properties of nervous system tissues and their integral role in neuroendocrine systems (see Table 1). Rather than provide a broad overview of these multiple approaches, we will focus in depth on methodologies currently in use in our laboratory, which incorporate single-unit analysis in combination with a detailed pharmacological assessment of identified neuronal receptor populations. This recording strategy can address the functional consequences of a given circulating hormone environment on identified biogenic amine-containing cells in brain regions with roles in the hypothalamic-pituitary-ovarian axis[17,18,19] and hippocampal-hypothalamic-pituitary-adrenal axis.[2,20]

TABLE 1
Reference Texts for Electrophysiological Recording Techniques

Authors/editors	Title	Publisher (year)
Brown, Kenneth T. and Flaming Dale G.	*Advanced Micropipette Techniques for Cell Physiology*	John Wiley & Sons (1992)
Kettenmann, Helmut and Grantyn, Rosemarie	*Practical Electrophysiological Methods: A Guide for In Vitro Studies in Vertebrate Neurobiology*	Wiley-Liss, Inc. (1992)
Oakley, Bruce and Schafer, Rollie	*Experimental Neurobiology: A Laboratory Manual*	The University of Michigan Press (1978)
Scherübl, Hans and Hescherler, Jürgen	*The Electrophysiology of Neuroendocrine Cells*	CRC Press (1996)
Sherman-Gold, Rivka	*The Axon Guide for Electrophysiology and Biophysics Laboratory Techniques*	Axon Instruments, Inc. (1993)
Stamford, J. A.	*Monitoring Neuronal Activity: A Practical Approach*	Oxford University Press (1992)

Note: These reference texts are recommended reading for additional technical information on cellular electrophysiological recording approaches and are readily available in most health science libraries. Please note: Not listed above are a wide range of excellent texts available on the theoretical aspects of nerve cell function, membrane biophysics and principles of electronics.

A. Extracellular Recording

Regardless of the type of recording preparation, several fundamental criteria must be addressed in order to successfully record the cell population of interest. A sufficiently stable preparation must be achieved which is typically accomplished by use of stereotaxic equipment. The remaining technical issues then focus on the detection of the signal of choice; this processes requires attention to (1) the characteristics of electrode, (2) preparation of the recording site, (3) amplification and discrimination of the signal, (4) identification of the waveform of interest, (5) collection of data and (6) histological confirmation of the recording site(s). Each of these issues will be discussed in detail below. The components of a typical recording set-up for *in vivo* extracellular recording is diagrammatically illustrated in Figure 1.

1. Electrodes

The electrodes used for extracellular recordings include single barrel and multi-barrel glass micropipettes. Several factors must be taken into consideration when preparing electrodes for electrophysiological experiments. For example, the tip diameter and final impedance of the micropipette is determined by the size of the cell to be recorded and the desired signal-to-noise ratio. The tip diameter should be less than that of the cell in order to obtain an adequate signal-to-noise ratio allowing for an easy and reproducible visualization of the desired waveform. Additionally, if the impedance is too low, cell firing may not be detectable. Alternatively, the electrode may become blocked and the signal will fail to be transmitted if the impedance is too high.

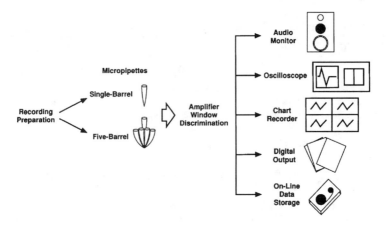

FIGURE 1

Schematic diagram of a typical *in vivo* extracellular recording preparation. The signal from the micropipette is obtained using a platinum lead placed in the recording barrel and connected to an amplifier and window discriminator. The amplifier is connected in series to an audio monitor, oscilloscope and chart recorder. The firing characteristics of the waveform frequency is digitized and stored with output to a digital printer and a personal computer.

Another consideration is the solution used to fill the electrode. For single barrel electrodes, 2 *M* NaCl or a 2% Pontamine Sky Blue (PSB) solution can be used. The 2 *M* NaCl solution may be advantageous over the 2% PSB if the recording session lasts for several hours. In this case, a charge may build up on the electrode which may accelerate the precipitation of the dye solution and produce a blockage of the electrode; the use of a concentrated salt solution alone will reduce this blockage. For multi-barrel micropipettes, the drug solutions used to fill the side barrels must be prepared at molar concentrations which will be stable in solution, and adjusted to a final pH. Additionally, it is important to use one side barrel for channel balancing to avoid the build up of electrical charge and thereby increase the noise output. The barrel which is used for channel balancing is routinely filled with a 2 to 4 *M* NaCl solution.[21]

When preparing to record from neurons of the dorsal raphe nucleus which will be examined for responsiveness to intravenous administration of a test compound or hormone, single barrel micropipettes (World Precision Instruments, Inc.; Sarasota, FL) are pulled to approximately a 5 mm tip length using a vertical electrode puller (Model PE-2, Narishigi; Tokyo, Japan). The electrode barrel is filled with 2 *M* NaCl or a 2% Pontamine Sky Blue solution for recording with a final impedance of 5 to 8 MΩ, determined *in vitro* in 0.9% NaCl at 60 Hz.[3]

Multi-barrel electrodes are used to microiontophoretically apply drug solutions directly to the extracellular environment of the cell. For example, five-barrel micropipettes (Model P-5, ASI Instruments, Inc.; Warren, MI) are pulled to approximately an 8 mm tip length and bevelled under a microscope to a total tip diameter of 6 to 10 μm for recordings of pyramidal neurons in the hippocampus. The central barrel is filled with a 2% Pontamine Sky Blue solution in 2*M* NaCl for recording with a final impedance of 1 to 4 MΩ. The remaining barrels are filled with the drugs of interest with one barrel reserved for channel balancing as described above.[3,4,5,22]

Single tungsten electrodes have also been effectively utilized for recording multi-unit activity in hypothalamic regions, including the magnocellular neurosecretory neurons in the paraventricular region[23] and in the arcuate nucleus.[14,15,16] While initial studies were conducted in anesthetized preparations, the use of metal electrodes has become highly refined and effective for long-term recordings in freely moving preparations in multiple mammalian species. Additionally, significant advances in microprocessor based technologies now facilitate simultaneous recordings from multiple individual units using arrays of tungsten electrodes (see Table 1). Certainly, with respect to analysis of circadian variations in neuronal sensitivity that may be modulated by endogenous variations in circulating steroid or hypothalamic releasing hormones (see Chapter 11), this will be a promising area for new developments in the field of neuroendocrinology.

2. Recording Site Preparation

Following preparation of the electrodes, the animal is anesthetized, the recording site and stereotaxic instrument prepared, and surgical tools are gathered. The lateral and vertical coordinates for the placement of the electrode in the brain region of interest are determined using an atlas specific for the species of interest. Atlases commonly used for investigations of the adult rat brain include those by König and Klippel,[24] Palkovits and Brownstein,[25] and Paxinos and Watson.[26] Using a hand-held drill equipped with a small diameter burr (3 mm), a small hole is drilled at the appropriate coordinates from bregma or other chosen landmark. Care should be taken to prepare a clean site, avoiding jagged edges of bone and damage to the surface of the dura. Should damage occur to surface blood vessels, bleeding must be controlled before the micropipette is lowered to the surface of the brain so as to avoid blockage of the electrode tip. Typically, the application of light pressure for 30 seconds will stop the bleeding; however, if this does not succeed, gelfoam (UpJohn Co.; Kalamazoo, MI) may be used. With bleeding controlled, the electrode is slowly lowered, the coordinates at the surface noted, and the micropipette manually lowered to a depth 1 mm above the vertical coordinates of the region of interest. The search for a cell is then performed using a microdrive (Model 640, David Kopf, Inc.; Tujunga, CA) to move the electrode in a step-wise fashion through the brain tissue in 1 μm increments.

3. Signal Amplification

The signal from the electrode is obtained through a platinum lead placed in the recording barrel and connected to a combination amplifier and window discriminator (WDR 420, Fintronics Inc.; Orange, CT). In order to achieve the needed signal-to-noise ratio, it is essential to minimize background noise with the use of a notch filter as well as low and high frequency cutoff filters. Typically, the amplifier output is displayed on an oscilloscope and the desired waveform is then isolated using a window discriminator. Functioning as a variable voltage window, the window discriminator commonly caps both the upper and the lower range of the spike deflection for a bipolar unit of interest. Care must be taken at this step to insure identification of a single- vs. multi-unit for single cell studies. Cell firing rates are permanently recorded as integrated rate histograms on a chart recorder (Recorder 220, Gould Instruments; Houston, TX) with a speed and sensitivity set according to cell firing rate characteristics. The number of spikes is recorded on a digital printer (Model DPP-Q7, Datel Inc.; Mansfield, MA).

4. Single-Unit Identification

Once a signal is obtained, single units must be distinguished from each other and from background noise. For example, single units in the dorsal raphe nucleus are identified by their characteristic slow, regular firing rates (1.0 to 3.0 Hz) and wide spike durations of 0.8 to 1.2 ms.[5,27] Additionally, an audio monitor (Grass AM8, Grass Instruments; Quincy, MA) is particularly helpful in discerning the presence of a single extracellularly recorded unit; often the cell type of interest can distinguished by "ear" prior to establishing an adequately amplified waveform. An inherent difficulty in the identification of spike activity from individual neurons is avoiding multi-unit recordings in regions with a dense population of neurons. Typical cells with this problem are found in the pyramidal layer of the hippocampus.[3,4,22]

An alternative approach for selection of target waveforms for recording is the application of electrical stimulation for antidromic identification. This approach has been particularly effective in elucidating functional aspects of identified neuronal input to the hypothalamic paraventricular[12] and the ventromedial[28] nuclei. The possible spread of uncontrolled electrical current, combined with the choice of placement of the stimulating electrode, are considerations in utilizing this methodology for elucidation of neuroendocrine circuitry.

5. Data Collection

Data may be collected and analyzed using a wide variety of methods. Using data output from the amplifier/window discriminator to the chart recorder, cell firing rate information can be obtained as an integrated rate histogram and the spike frequency counted on a digital printer for permanent storage. Alternatively, a range of analogue to digital conversion software, designed for compatibility with personal computers, are available for capture of data as well as on- and off-line analysis of data. We will not review the advantages and limitations of the numerous commercially available systems but encourage the reader to directly consult the manufacturers for further information (see Table 1).

6. Histological Confirmation of the Recording Site

Recording sites in brain tissues are readily verified by post-mortem histological examination following ejection of a dye or by production of an electrolytic lesion at the electrode tip. Confirmation of the placement of each recording site is necessary to insure that the results accurately reflect the chosen cell population.

With the electrode placed at the site of the last recorded unit, a current of 20 μA is passed through the recording barrel of the electrode for a range of 5 to 40 min using a bipolar constant current source (VL-1200d, Fintronics; Orange, CT). The brain is removed and stored in 10% buffered formalin. Visualization of the dye spot or lesion produced by current ejection should agree with the original electrode placement, stereotaxic coordinates chosen, and be consistent with the waveforms identified during the recording session. Recognizing that diverse cell types (and sizes) are present within a brain region, such as the ventromedial nucleus of the hypothalamus[28] or the midbrain dorsal raphe nucleus,[27] is vital. The final identification of a cell type of interest must reflect all aspects (stereotaxic location, waveform characteristics, cell firing pattern, neuronal responsiveness, histological location) of the data collected.

B. Evoked-Potential Recording

In addition to the single-unit recording procedures describe above, extracellular recording techniques can be used for studies of the effects of hormone on electrical stimulation-induced evoked potentials. This approach is now most commonly used with *in vitro* recording preparations to address issues of firing frequency and latency. For example, estradiol has been demonstrated to increase the extracellularly recorded hippocampal pyramidal cell activity when electrically evoked in the region of the Schaffer collaterals.[29] This approach can be combined with *in vivo* recording approaches to identify changes in synaptic events as regulated by the high and low levels of a given circulating hormone.

C. Intracellular Recording

In vivo recording preparations are seldom utilized for intracellular recordings of central neurons as it is extremely difficult to provide the requisite degree of mechanical stability needed for the recordings. The use of the brain slice or isolated cell culture is a feasible approach and allows for the direct testing of a hormone on identified pre-or post-synaptic membrane events (see Chapter 6).

IV. Neuropharmacologic Characterization of Cellular Responses

Extracellular recording techniques are an excellent approach to characterize the effects of a hormone on a given neuronal target. Combined with systemic administration into the general circulation or a discreet placement of the compound of interest onto the target cell being recorded, detailed information on the pharmacology of a given hormone and/or modulator can be obtained. These approaches for drug administration can be highly useful as tools for defining the neuronal sensitivity of a pharmacologically defined cell population recorded in the brain.

A. Systemic Drug Infusion Techniques

Systemic administration of drugs can be accomplished by intravenous infusion, intramuscular injection or subcutaneous injection or pellet implant. Typically, to determine an acute effect of a test compound on the firing of a novel compound, the compound is dissolved in an aqueous solution and administered intravenously in a cumulative log dose manner.[22] This approach allows for direct comparison of compounds, often a novel drug to a known standard.[3,22] The comparison of neuronal responses obtained in animals recorded at different ages across the lifespan is also feasible.[3,4,19]

It is important to be aware of the solubility of the compound to insure that it can be administered in an intravenous solution. Likewise, another consideration with the systemic administration of hormones is whether the compound administered will cross the blood-brain barrier and enter the central nervous system. It is important to recognize that should a lack of effect on the target cell be observed for a novel compound, the result may reflect a range of pharmacokinetic issues, including the lipid solubility, diffusion rate, and metabolism of the compound.

B. Micro-Iontophoretic Drug Application Techniques

Micro-iontophoresis involves the ejection of ionized solutions from micropipettes.[30] The fundamental principle of iontophoresis is that opposite charges attract and like charges repel. Therefore, to retain a solution in a given barrel of an electrode an opposite charge is applied to the solution filling the barrel. Alternatively, a charge identical to that of the drug is applied to the interior of the barrel during the process of drug ejection.

In microiontophoresis, the transport number of a substance is used to calculate the amount of drug released from the electrode barrel. Using this number, the dose-response characteristics of different compounds can be compared. The transport number of a compound depends on its solubility, dissociation, polarity and external medium. Theoretically, the transport number is proportional to the amount of current applied, such that the greater the current applied the more drug is released. Care must be taken to investigate drug-mediated responses that are obtained in the linear portion of the current response curve.

One advantage of iontophoresis is that small amounts of solution can be locally applied to a single neuron. Therefore, the effects of a drug can be examined in a specific brain region without affecting other brain regions or peripheral physiological systems. This technique also bypasses the blood-brain barrier which normally functions to inhibit the passage of charged compounds into the brain and can establish the direct effects of a novel compound or endocrine factor on neural tissue. Furthermore, the use of multi-barrel pipettes allows one to test and compare the effects of several compounds on the same neuron.

Although the transport number indicates the total number of ions ejected, a limitation of microiontophoresis is that the actual volume of drug ejected cannot be determined. The routine application of a randomized range of currents for a given compound will, however, insure that the observed excitation and/or inhibition are independent of the volume of drug ejected. Other technical problems that may be encountered with iontophoresis include blockage of electrode barrels by brain tissue, blood clots and/or charge capacitance resultant from the drug solution. These problems can be minimized by careful preparation of the recording site and consistent electrode preparation for every recording session.

Evaluation of the direct effects of steroid hormones, including estrogen, progesterone and corticosteroids, are limited due to the lack of ionic charge and limited solubility in water of these compounds. Derivatives of steroid hormones that demonstrate an enhanced solubility in polar solutions are becoming commercially available and, therefore, these compounds may be utilized in microiontophoretic studies to evaluate their direct effects in the nervous system. An alternative design for application of steroids is the use of pressure ejection techniques. While uncontrolled diffusion of compounds is a complication of this approach, pressure ejection has been effectively used for analysis of the effects of estrogen.[31]

Analysis and interpretation of microiontophorectic drug application may include a variety of measurements, ranging from determination of the frequency of spontaneous cell firing rates to demonstrations of selective pharmacological blockage of a current-dependent drug-induced response. Commonly, the inhibition of cell firing produced by microiontophoretic drug application is assessed using an IT_{50} value. As defined by de Montigny and Aghajanian,[32] the IT_{50} value is equal to the current in nA multiplied by the time in seconds required to obtain a 50% decrease in firing rate. This parameter is expressed in nanocoloumbs and is an index of postsynaptic neuronal sensitivity. The recovery period following application of a drug is also a useful physiological index of cell function. For example, the recovery phase following inhibition of cell firing may be examined using the RT_{50} value which is an index of the ability of a presynaptically located transporter to remove neurotransmitter from the synaptic cleft.[33] The RT_{50} value is defined as the time in seconds required for a cell to recover to 50% of the mean baseline firing rate immediately following drug microiontophoresis.

It is important to recognize that the application and comparison of compounds by microiontophoresis is a semi-quantitative technique involving analysis of ion migration. Many technical factors have been identified to influence the time-course and intensity of a neuronal response during microiontophoresis including the transport number of the drug, type of electrode, the duration of iontophoresis, as well as the intensity and duration of the retaining current.[21,34,35]

V. Future Directions

Many questions remain regarding the function and regulation of neuroendocrine systems that are well suited for investigation by *in vivo* recording techniques. Clearly this technology will continue to be vital in establishing the neurocircuitry of neuroendocrine systems. We are likely to see new developments in the near future that will permit the simultaneous recording of identified neurons at multiple targets simultaneously. For example, such an approach might be able to investigate the effects of stress-induced elevation in corticosterone on serotonin-mediated responses recorded in the dorsal raphe nucleus and post-synaptic target tissues recorded in the forebrain (including the hippocampus and paraventricular nucleus) simultaneously. Likewise, these approaches will be invaluable as we elucidate the dynamics of neuroendocrine systems across the lifespan. Understanding the function of an identified neuron in response to multiple endocrine states will help us to better understand reproductive aging and age-dependent losses in ability to adapt to stress and other age-associated neuroendocrine events.

VI. Conclusions

This chapter has addressed the application of *in vivo* electrophysiological recording approaches to the study of neuroendocrine systems. We have emphasized the uses of varied recording preparations and covered the technical details of several recording techniques as well as their application to the study of hormone-mediated responses in various brain regions. As summarized in Table 2, the cellular electrophysiological response profiles to a variety of hormones have been investigated and provide a solid basis for current studies of nervous and endocrine system interaction. These techniques offer the neuroendocrinologist an active view of synaptic activity that is inherent in receptor/effector coupling in response the dynamics of hormone secretion.

Acknowledgments

Support for this work was provided by U.S.P.H.S. Grant PO1 AG10514 (J.M.L.) and T32 AG00048 (J.E.S.) awarded by the National Institute on Aging (J.M.L.) and the Department of Pharmacology, Pennsylvania State University College of Medicine.

TABLE 2
Neuroendocrine Studies Utilizing Electrophysiological Recording Techniques

Hormones	References (year)
Estrogen	Arentsen, M. and Lakoski, J. M. (1993)[17]
	Chido, L. A. and Caggiula, A. R. (1980)[36]
	Kelly, M. J. et al., (1976, 1977)[37,38]
	Lakoski, J. M. (1997)[39]
	Marcus, E. M. et al., (1966)[40]
	Pfaff, D. W. (1980)[28]
	Poulain, P. and Carette, B. (1981)[31]
	Smith, S. S. et al., (1987, 1988)[41,42]
	Terasawa, E. and Timiras, P. S. (1968)[43]
	Teyler, T. J. et al., (1980)[29]
	Yagi, K. (1973)[44]
Progesterone	Smith, S. S. et al., (1987)[42]
	Terasawa, E. and Sawyer, C. H. (1970)[43]
Corticosteroids	Joëls, M. et al., (1994)[20]
	Kelly, M. J. et al., (1977)[45]
	Pfaff et al., (1971)[46]
Gonadotropin-Releasing Hormone	Hayward, J. N. (1977)[23]
(GnRH, LH-RH)	Knobil, E. (1990)[14]
	Silverman, A. J. et al., (1986)[15]
	Wilson, R. C. et al., (1984)[16]
Corticotropin-Releasing Factor	Eberly, L. B. et al., (1983)[47]
(CRF)	Siggins, G. R. (1990)[48]
	Valentino, R. J. (1990)[6]
	Valentino, R. J. et al., (1983)[7]
Oxytocin and Vasopressin	Renaud, L. P. and Bourque, C. W. (1991)[12]
	Wakerley, J. B. and Ingram, C. D. (1993)[13]

Note: Provided as a guide for additional reading, these references are a partial listing of *in vivo* cellular electrophysiological studies of hormones which serve as key signaling molecules in several neuroendocrine systems. Note: Electrophysiological studies of classical neurotransmitters (e.g. serotonin, dopamine, norepinephrine, amino acids, opioids) which also serve as signaling molecules in multiple neuroendocrine systems are not listed.

References

1. Lakoski, J. M., Keck, B. J. and Dugar, A., Neurochemical lesions: tools for functional assessment of serotonin neuronal systems, in *Paradigms of Neural Injury, Methods in Neuroscience, Vol. 30,* J. R. Perez-Polo, Ed., Academic Press, New York, 1996, 115.

2. Maines, L. W., Keck, B. J. and Lakoski, J. M., Serotonin 5-HT$_{1A}$ receptors are differentially regulated by corticosterone across brain regions in an age-dependent manner, *Soc. Neurosci. Abstr.,* 22, 1889, 1996.

3. Smith, J. E. and Lakoski, J. M., Electrophysiological study of the effects of the reuptake inhibitor duloxetine on serotonergic responses in the aging hippocampus, *Pharmacology,* 55, 66, 1997.

4. Dugar, A. and Lakoski, J. M., Serotonergic function of aging hippocampal CA3 pyramidal neurons: Electrophysiological assessment following administration of 5,7-dihydroxytryptamine in the fimbria-fornix and cingulum bundle, *J. Neurosci. Res.,* 47, 58, 1997.

5. Lakoski, J. M. and Aghajanian, G. K., Effects of histamine, H1- and H2-receptor antagonists on the activity of serotonergic neurons in the dorsal raphe nucleus, *J. Pharmacol. Exp. Ther.,* 227, 517, 1983.

6. Valentino, R. J., Effects of CRF on spontaneous and sensory-evoked activity of locus coeruleus neurons, in *Corticotropin-Releasing Factor: Basic and Clinical Studies of a Neuropeptide,* E. B. De Souza and C. B. Nemeroff, Eds., CRC Press, Boca Raton, FL, 1990, 217.

7. Valentino, R. J., Foote, S. L. and Aston-Jones, G., Corticotropin-releasing factor activates noradrenergic neurons of the locus coeruleus, *Brain Res.,* 270, 263, 1983.

8. Wise, P. M., Aging of the female reproductive system, *Rev. Biol. Res. Aging,* 1, 195, 1983.

9. Baghdoyan, H. A., Spotts, J. L. and Snyder, S. G., Simultaneous pontine and basal forebrain microinjections of carbachol suppress REM sleep, *J. Neurosci.,* 13, 229, 1993.

10. Ito, K. and McCarley, R. W., Alterations in membrane potential and excitability of cat medial pontine reticular formation neurons during changes in naturally occurring sleep-wake states, *Brain Res.,* 292, 169, 1984.

11. Leonard, T. O. and Lydic, R., Nitric oxide synthase inhibition decreases pontine acetylcholine release, *Neuroreport,* 6, 1525, 1995.

12. Renaud, L. P. and Bourque, C. W., Neurophysiology and neuropharmacology of hypothalamic magnocellular neurons secreting vasopressin and oxytocin, *Prog. Neurobiol.,* 36, 131, 1991.

13. Wakerley, J. B. and Ingram, C. D., Synchronisation of bursting in hypothalamic oxytocin neurones: possible coordinating mechanisms, *News Physiol. Sci.,* 8, 129, 1993.

14. Knobil, E., Electrophysiologic approaches to the hypothalamic GnRH pulse generator, in *Neuroendocrine Regulation of Reproduction,* S. S. C. Yen and W. W. Vale, Eds., Serono Symposia, 1990, 3.

15. Silverman, A.-J., Wilson, R., Kesner, J. S. and Knobil, E., Hypothalamic localization of multiunit electrical activity associated with pulsatile LH release in the rhesus monkey, *Neuroendocrinology,* 44, 168, 1986.

16. Wilson, R. C., Kesner, J. S., Kaufmann, J.-M., Uemura, T., Akema, T. and Knobil, E., Central electrophysiologic correlates of pulsatile lutenizing hormone secretion in the rhesus monkey, *Neuroendocrinology,* 39, 256, 1984.

17. Arensten, M. I. and Lakoski, J. M., Cellular electrophysiology of 5-HT$_{1A}$-mediated responses in CA1 and CA3 hippocampal subfields with aging, *Soc. Neurosci. Abstr.,* 1993.

18. Lakoski, J. M., Neuroendocrinology of aging at the cellular level: Membranes to neural circuits, *Neurobiol. Aging,* 15, 519, 1994.

19. Lakoski, J. M., Cellular electrophysiological approaches to the central regulation of female reproductive aging, in *Neural Control of Reproductive Function,* J. M. Lakoski, J. R. Perez-Polo and D. A. Rassin, Eds., Alan R. Liss, New York, 1989, 209.

20. Joëls, M., Hesen, W., Karst, H. and de Kloet, E. R., Steroid and electrical activity in the brain, *J. Steroid Biochem. Mol. Biol.,* 49, 391, 1994.

21. Salmoiraghi, G. C. and Weight, F., Micromethods in neuropharmacology: an approach to the study of anesthestics, *Anesthesiology,* 28, 54, 1967.

22. Smith, J. E. and Lakoski, J. M., Electrophysiological effects of fluoxetine and duloxetine in the dorsal raphe nucleus and hippocampus, *Eur. J. Pharmacol.,* 323, 69, 1997.

23. Hayward, J. N., Functional and morphological aspects of hypothalamic neurons, *Physiol. Rev.,* 57, 574, 1977.

24. König, J. F. R. and Klippel, R. A., *The Rat Brain: A Stereotaxic Atlas,* Williams and Wilkins, Baltimore, 1963.

25. Palkovits, M. and Brownstein, M. J., *Maps and Guide to Microdissection of the Rat Brain,* Elsevier Science, 1988.

26. Paxinos, G. and Watson, C., *The Rat Brain in Stereotaxic Coordinates,* Academic Press, New York, 1986.

27. Aghajanian, G. K., Feedback regulation of central monoaminergic neurons: Evidence from single cell recording studies, in *Essays in Neurochemistry and Neuropharmacology,* M. B. H. Youdim, D. F. Sharman, W. Lovenberg and J. R. Lagnado, Eds., John Wiley & Sons, New York, 1978, 1.

28. Pfaff, D. W., *Estrogens and Brain Function,* Springer-Verlag, New York, 1980.

29. Teyler, T. J., Vardaris, R. M., Lewis, D. and Rawitch, A. B., Gonadal steroids: effects on excitability of hippocampal pyramidal cells, *Science,* 209, 1017, 1980.

30. Hicks, T. P., The history and development of microiontophoresis in experimental neurobiology, *Prog. Neurobiol.,* 22, 185, 1984.

31. Poulain, P. and Carette, B., Pressure ejection of drugs on single neurons in vivo: technical considerations and applications to the study of estradiol effects, *Brain Res. Bull.,* 1, 33, 1981.

32. de Montigny, C. and Aghajanian, G. K., Preferential action of 5-methoxydimethyltryptamine on presynaptic serotonin receptors: a comparative iontophoretic study with LSD and 5-HT, *Neuropharmacology,* 16, 811, 1977.

33. de Montigny, C., Wang, R. Y., Reader, T. A. and Aghajanian, G. K., Monoaminergic denervation of the rat hippocampus: Microiontophoretic studies on pre- and postsynaptic supersensitivity to norepinephrine and serotonin, *Brain Res.,* 200, 363, 1980.

34. Bloom, F. E., Siggins, G. R. and Henriksen, S. J., Electrophysiologic assessment of receptor changes following chronic drug treatment, *Fed. Proc.,* 40, 166, 1981.

35. Curtis, D. R. and Eccles, R. M., The effect of diffusional barriers upon the pharmacology of cells within the central nervous system, *J. Physiol. (London),* 141, 446, 1958.

36. Chiodo, L. A. and Caggula, A. R., Alterations in basal firing rat and autoreceptor sensitivity of dopamine neurons in the substantia nigra following acute and extended exposure to estrogen, *Eur. J. Pharm.,* 67, 165, 1980.

37. Kelly, M. J., Moss, R. L. and Dudley, C. A., The specificity of the response of preoptic-septal area neurons to estrogen: 17α-estradiol versus 17β-estradiol and the response of extrahypothalamic neurons, *Exp. Brain Res.,* 30, 43, 1977.

38. Kelly, M. J., Moss, R. L. and Dudley, C. A., Differential sensitivity of preoptic-septal neurons to microelectrophoresed estrogen during the estrous cycle, *Brain Res.,* 114, 152, 1976.

39. Lakoski, J. M., Estrogen and the cellular physiology of serotonergic neuronal systems: a key to the aging brain? *Biol. Psychiatry,* 42, 7S, 1997.

40. Marcus, E. M., Watson, C. W. and Goldman, P. L., Effects of steroids on cerebral electrical activity, *Arch. Neurol.,* 15, 521, 1966.

41. Smith, S. S., Waterhouse, B. D. and Woodward, D. J., Sex steroid effects on extrahypothalamic CNS. I. Estrogen augments neuronal responsiveness to iontophoretically applied glutamate in the cerebellum, *Brain Res.,* 422, 40, 1987.

42. Smith, S. S., Waterhouse, B. D., Chapin, J. K. and Woodward, D. J., Progesterone alters GABA and glutamate responsiveness: a possible mechanism for its anxiolytic action, *Brain Res.,* 400, 353, 1987.

43. Terasawa, E. and Timaras, P. S., Electrical activity during the estrous cycle of the rat: cyclic changes in limbic structures, *Endocrinology,* 86, 207, 1968.

44. Yagi, K., Changes in firing rates of single preoptic and hypothalamic units following an intravenous administration of estrogen in the castrated female rat, *Brain Res.,* 53, 343, 1973.

45. Kelly, M. J., Moss, R. L. and Dudley, C. A., The effects of microiontophoretically applied estrogen, cortisol and acetylcholine on medial preoptic septal unit activity throughout the estrous cycle of the female rat, *Exp. Brian Res.,* 30, 53, 1977.

46. Pfaff, D. W., Silva, M. T. A. and Weiss, J. M., Telemetered recording of hormone effects on hippocampal neurons, *Science,* 171, 394, 1971.

47. Eberly, L. B., Dudley, C. A. and Moss, R. L., Iontophoretic mapping of corticotropin-releasing factor (CRF) sensitive neurons in the rat forebrain, *Peptides,* 4, 837, 1983.

48. Siggins, G. R., Electrophysiology of cortiocotropin-releasing factor in nervous tissue, in *Corticotropin-Releasing Factor: Basic and Clinical Studies of a Neuropeptide,* E. B. De Souza and C. B. Nemeroff, Eds., CRC Press, Boca Raton, FL, 1990, 205.

Chapter 8

Antisense Oligonucleotides as Tools in Neuroendocrinology

Mariana Morris and Ping Li

Contents

Antisense oligodeoxynucleotides (ODNs) are increasingly used as tools to probe neurobiological problems.[1-3] They are promoted as alternatives to pharmacological antagonists with the major advantage being that they can be used to specifically target prohormones, receptors and intracellular mediators. This is especially important in situations in which the targets are members of superfamilies, making it difficult to construct specific antagonists, or when the agonists/antagonists are unknown. In many cases antisense have been shown to exert similar effects as antagonists, modulating behavior, cardiovascular parameters and neurosecretion.[4-14] Nonetheless, there are questions regarding their mechanisms of action, specificity and possible toxicity. This chapter will focus on the use of antisense ODNs in the study of the central nervous system (CNS), including discussion of their advantages and disadvantages.

I. Antisense Oligodeoxynucleotide Uptake and Stability

It is well known that nucleases present in cells and plasma degrade DNA. Thus, one of the first objectives in the study and use of antisense ODNs was to develop forms which are not readily degraded, providing a longer half/life and presumably greater efficacy. The most widely used nuclease-resistant ODN is the phosphorothioate (PS) form in which a sulfur atom replaces one of the non-bridging oxygen atoms at each internucleoside phosphorus atom. A comparison of the PS to the naturally occurring phosphodiester (PO) form reveals that modification increases the half/life.[15] However, the rate of degradation is dependent on the biological compartment with increased stability of both forms noted in brain, cerebrospinal fluid and neuronal cultures as compared to plasma, probably because of lower nuclease activity in the CNS.[6,16,17] There is rapid accumulation of ODNs in the cytoplasm (within 5 to 15 min), with greater uptake in neurons than astrocytes.[18] The unmodified forms showed a more site-localized spread as compared to the PS forms which produced a diffuse spread with evidence for an inflammatory reaction.[19] Even though PO-ODNs are more readily degraded than PS-ODNs, there are data showing the presence of PO-antisense in cells up to 24 h after administration and evidence for the formation of stable duplexes.[6,20]

II. Mechanisms of Antisense Action

The basic idea behind the use of antisense ODNs is that a sequence complimentary to the mRNA will bind via hydrogen bonding base-pairing and influence translation. The optimal length is 15-25 base pairs with the sequence usually targeted toward the 5' coding region. It has been postulated that ODNs act to inhibit the ribosomal read-through of the mRNA transcript and/or act as a substrate for RNase H which degrades the RNA-DNA hybrid.[21,22] With each scenario the endpoint is a reduction in gene product, the neuropeptide or its receptor (Figure 1). In cell free and single cell systems, RNase H activity may mediate antisense effects; however, in the CNS there is more evidence to support a role for message translation. Most studies reveal that antisense treatment has little effect on mRNA levels and in some cases may actually increase concentrations.[5,6,11,14,15,23-26] Landgraf and colleagues tested the effect of antisense and sense ODN to vasopressin V-1 receptor mRNA and reported that the antisense increased mRNA levels while the sense construct reduced them.[5] Likewise, increases were seen for corticotropin releasing hormone and glucocorticoid and mineralocorticoid receptor mRNA after antisense treatment.[26] The idea emerges that there are interactions between translation and transcription, a feedback loop within the neuron, acting to counteregulate the gene product deficiency. Thus, there are opposing actions within the antisense-affected cell, arrest of translation produced by the antisense-mRNA hybrid and a resulting increase in mRNA to

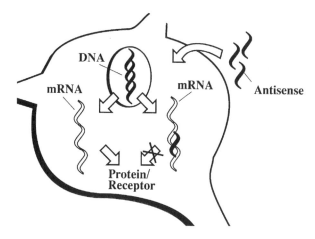

FIGURE 1
Hypothetical scheme for antisense action.

promote protein synthesis. This illustrates the complex interrelationship between mRNA, protein and the time course of turnover, making it difficult to arrive at testable mechanistic approaches.

III. Effectiveness of Antisense Treatment

One difficulty in the determination of antisense effectiveness is the selection of the appropriate endpoint for study. Since antisense presumably acts to alter the synthesis cascade, one could measure changes in mRNA, tissue proteins, receptors or biosynthetic turnover. Changes in peptide content (direct measurement or immunocytochemical staining) and receptor density (peptides and neurotransmitters) have been reported after acute and chronic antisense treatment.[5,6,9,12,13,14,23,27-32] We observed a decrease in paraventricular (PVN) oxytocin content but no change in peptide stores in the neural lobe after PVN injection of antisense to oxytocin mRNA.[9] Using a similar antisense with immunochemical staining as the criteria, divergent results were reported, either a decrease[25] or no change[4] in nuclear staining. Antisense to neuropeptide Y (NPY) blocked the increase in NPY produced by steroid stimulation, an effect attributed to inhibition of peptide synthesis.[27] Likewise, antisense to other peptide hormones decreased cellular immunostaining.[6,25,30] However, a direct link between antisense administration and peptide biosynthesis has not been established and is impossible to verify by measurement of tissue content or immunocytochemical staining. A more detailed and direct analysis is required with evaluation of the effect of antisense treatment on the time course of biosynthesis and peptide processing.

    With regard to receptors, there is uniform agreement that antisense treatment lowers receptor density. This is seen for angiotensin, vasopressin, oxytocin, dopamine and other receptors with specificity noted for the receptor subtype. For example,

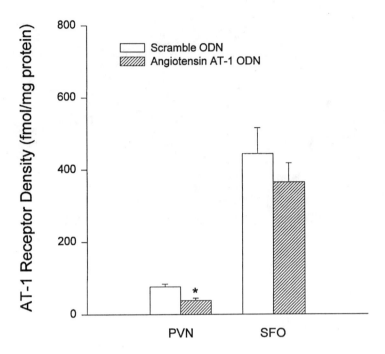

FIGURE 2

There was a 40% decrease in PVN AT-1 receptors 20 hr after direct PVN injection of AT-1 AS ODN, compared with scramble ODN in Tg rats. In contrast, there were no differences in SFO AT-1 receptors.

angiotensin AT-1 antisense decreased AT-1, but not AT-2 receptors;[29] dopamine D-2 antisense lowered D-2 receptors and not D-1 receptors;[14] and antisense to NPY-1 receptors reduced the NPY-1 subtype without affecting NPY-2 receptors.[13] Likewise, there are localized changes in receptor density seen after tissue injection of antisense.[12] When antisense to angiotensin AT-1 receptor was injected directly into the PVN, there was a reduction in PVN AT-1 receptors, but no alteration in subfornical AT-1 receptors (Figure 2). Upon comparison, cerebroventricular antisense injection resulted in widespread receptor changes, most notably in areas close to the ventricular surface.[2]

Even though receptors and peptides are reduced by antisense treatment, the time course and the degree of changes do not correlate well with functional deficits. Receptor depletion is always incomplete, with changes as little as 10% even with chronic treatment.[5,12,13,14,28,29,32] This occurs in the face of more profound alterations in behavioral, endocrine and cardiovascular responses. Furthermore, recovery of function is seen even when receptors continue to be depressed. For example, we found that the antisense to the angiotensin AT-1 receptor produced a rapid and short-lived reduction in blood pressure with a longer term change in receptors (40% reduction 20 h after antisense injection).[12] Investigators have theorized that the dichotomy may relate to rapid changes in a pool of functionally active receptors or changes in a releasable peptide pool. Indeed, a coupling of antagonist and antisense treatments for dopamine D-2 receptors suggested that antisense inhibits the synthesis of a sub-population of dopamine receptors responsible for stereotypical behavior.[31] One should also consider that there may be non-uniform uptake of antisense and regional or subcellular changes which are difficult to measure. To circumvent the problem of uptake and localization, Berrow and colleagues micro-injected antisense into cultured dorsal root ganglion neurons and measured the time course of the changes in protein levels and function.[33] Injection of antisense to the β subunit of the calcium channel produced almost a 100% depletion of immunoreactivity coupled with a reduction in calcium currents. The single antisense injection produced a long-term change with recovery beginning more than 4 days after injection. These results verify that part of the problem in interpreting the concomitant changes in proteins, receptors and function in whole animals or tissues is a lack of discrimination of the techniques, i.e., evaluation of regional rather than cellular changes.

IV. Antisense Oligodeoxynucleotide Toxicity and Nonspecificity

When tissue culture studies have employed ODNs, either directly injected into cells or delivered in the media, there is evidence for reversibility of effect without toxicity.[33,34] Likewise, when using acute intracerebral injections with relatively short survival times, tissue damage was not observed.[4,9,12] Chronic infusions are reported to provide a better method of antisense delivery; however, there is controversy as to whether this method causes tissue damage.[5,35,36] There are an increasing number of studies which report behavioral alterations, tissue damage and immune responses to antisense therapy. Skutella et al. reported that ventricular injection of antisense caused generalized demise with fever and piloerection.[37] Lucion and Morris found similar, toxic effects after two or more ventricular injections of antisense ODN, resulting in a significant degree of morbidity (unpublished observations). A cumulative toxic reaction was seen after multiple injections of PS-antisense or sense ODN (cFos) into the amygdala.[38] These investigators compared the effect of different injection schedules and found that tissue damage was correlated with the number and spacing of injections. Large lesions with glial infiltration were noted even when injections were spaced as much as 3 days apart while a single injection or injections given with a 5-day interval caused no damage. Similarly, Karle et al. showed extensive brain damage after chronic parenchymal infusion of PS-antisense to the GABAa receptor.[36] They attributed the neurotoxic damage to changes in GABA receptors, a questionable conclusion given the level of tissue damage. Finally, there are studies showing inflammatory and stress responses to PS-antisense injection.[19,37] Part of the discrepancies with these reports may be the chemical form of the antisense used, the route of administration or the dose. In general, the PS form was reported to be more neurotoxic than the unmodified form,[17,23] perhaps because of its avidity for cellular proteins.[39-41] The intraventricular route may cause more problems than direct tissue injection, possibly because of its spread to multiple regions and finally higher doses and multiple injections or infusions are likely to cause more non-specific and toxic effects.

While there is strong evidence that antisense ODNs are effective in modulating genetic expression, there are also data to suggest that there are interactions with other proteins and enzymes, non-sequence dependent effects and tissue toxicity (discussed above). The actions of the modified forms, particularly at higher doses, may be independent of base sequence, resulting in changes in other mRNA species, binding to essential cell components or other toxic effects.[39,40,42] For example, PS ODNs inhibited human RNase H and DNA polymerase activity, effects which were independent of sequence, but dependent on the number of phosphorothioate linkages.[43] Recent studies showed that PS ODNs bind to heparin binding growth factors, basic fibroblast growth factor and platelet-derived growth factor with a higher affinity for the PS forms than the PO.[40] It is suggested that these protein interactions are an important part of the mechanism by which ODNs modulate vascular proliferation.[41] Indeed the early enthusiasm for the use of antisense ODNs to proto-oncogenes to inhibit restenosis has been tempered by the findings that the action may be independent of nucleotide sequence.

V. Importance of Appropriate Controls for Antisense Experiments

A critical issue in the use of antisense ODNs in neuroendocrine studies is the inclusion of extensive and appropriate controls. In the study of behavior, neurosecretion, blood pressure or other physiological endpoints, any non-specific effects could alter the conclusions and lead to a misinterpretation of the findings. For example, Skutella and colleagues examined the effect of acute injection of antisense and sense ODN to corticotropin releasing hormone mRNA and found that both produced an increase in corticosteroid secretion, indicative of a stress response.[37] However, studies of oxytocin, another stress-responsive hormone, showed no evidence of non-specific ODN effects.[9,44] Changes in behavior have also been noted after antisense treatment; for example, alterations in sleep patterns after sense ODN to CFOS,[45] barrel rolling after scramble or dopamine D-2 antisense,[14] and decrease in drinking after high doses of missense ODN.[7] The scientific consensus is that studies employing antisense techniques must employ a variety of controls as well as endpoints for measurement. Chemical controls are the first to be considered in the design of antisense studies. Three types of chemical controls that are normally used are (1) scramble or random ODNs with a similar base composition as the antisense (2) sense ODNs with a structure complementary to the antisense and (3) mismatch in which 2 to 4 bases in the antisense are replaced. The antisense may be the unmodified PO form, sulfur modified PS or other modifications such as the 3'-3' end-inverted internucleotidic linkage. Scramble controls have been criticized because they may not be taken up within the same intracellular compartment as the antisense. Problems with sense controls have been reported since in some studies they have

TABLE 1
Control Groups Used for the Study of Salt-Induced Hypertension in the mRen-2 Transgenic (Tg) Rat

Group	Treatment	PVN-ODN
Tg	Salt	AT-1
Tg	Salt	Scramble
Tg	Water	AT-1
Tg	Water	Scramble
Control	Salt	AT-1
Control	Water	Scramble

Note: Male rats with chronic arterial catheters and PVN guide cannula were given 2% NaCl or water to drink for 4 days. Angiotensin AT-1 AS or scramble (1.8 μg) was injected into the PVN with measurement of blood pressure and heart rate.

similar effects as antisense, perhaps because of their association with nuclear DNA.[3] Mismatch of 3 to 5 base pairs offers a useful control since it is closely associated with the antisense but presumably less effective. Indeed in most cases mismatch sequences produce little effect; however, in one study of cFos, a mismatch of 3 out of 15 bases produced the same effect as the antisense.[38] Another useful approach in the study of antisense is the inclusion of physiological controls, animals under different physiological conditions. For example, to study the role of angiotensin AT-1 receptors in the maintenance of salt-induced hypertension, we tested the effect of PVN injection of unmodified antisense to the AT-1 receptor or scramble into mRen-2 Tg (Tg) or control rats receiving 2% NaCl or water to drink.[12] The groups and their treatment are outlined in Table 1. The results demonstrate that the AT-1 antisense had a depressor effect only in salt loaded Tg rats, not in the control rats or in Tg rats consuming water (Figure 3). The scramble sequence had no effect in any of the groups (Figure 4). Thus, the effectiveness of the antisense treatment was dependent on the genetic group and the physiological status of the animal. This type of protocol combines chemical controls (antisense vs. scramble), genetic controls (Tg vs. Control) and physiological controls (salt vs. water consumption). Specificity for a genetic model was also shown in studies by Gyurko et al. in which they used the SHR model of hypertension and showed that antisense to angiotensinogen or AT-1 receptors decreased blood pressure in hypertensive but not normotensive animals.[32]

Another type of control is the use of antisense to different but similar mRNA or the measurement of several endpoints (i.e, receptors, behavior, neurosecretion). For example, in our studies of stress-induced tachycardia, we compared the effect of antisense to two very similar hormones, vasopressin and oxytocin.[9] Stress-induced tachycardia was abolished by oxytocin antisense with no effect of vasopressin or scramble. Similarly, Neumann et al. showed that oxytocin antisense was specific for suckling-induced milk ejection and neuronal electrophysiological responses.[4,46] Comparison of the secretion of the two hormones combined with different antisense treatments is another mode of differentiation. Using a perfused hypothalamic explant model, we demonstrated that oxytocin antisense affected only oxytocin secretion

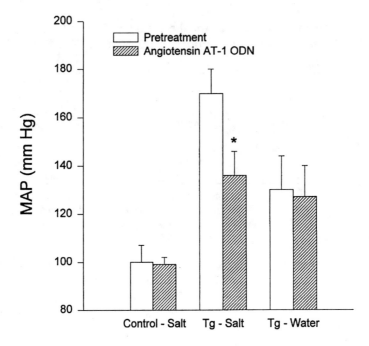

FIGURE 3
PVN injection of AT-1 AS ODN produced a rapid decrease in MAP of 34 mmHg in salt-loaded Tg rats at 3 hr after injection. There was to difference in either salt loaded control or water-replete Tg rats.

and vice versa for vasopressin, while scramble ODN had no significant effects.[11] Finally, one could include other non-antisense methods as a comparison, i.e, antagonist treatment or brain lesions. In studies of maternal behavior, Lucion and colleagues compared the effect of PVN lesions to direct PVN injection of oxytocin antisense and found that both gave similar results.[10] The key issue is that there are many caveats in the use of antisense ODNs to study physiological mechanisms. Paramount in the design of these studies is the inclusion of sufficient controls, including different chemical sequences, chemical type, parameters measured and physiological models.

VI. Other Aspects of Antisense Action

Some intriguing aspects of antisense function which may hold clues to their mechanism of action are (1) their rapid onset of action and (2) their lack of effect on basal status and attenuation of stimulated responses. This is seen for behavior, cardiovascular parameters, neurophysiological activity, endocrine secretion and others (Table 2). In our studies of cardiovascular control, antisense to oxytocin, vasopressin, or angiotensin AT-1 receptors had no effect on basal blood pressure or heart rate.[9,12] Only when the animals were stressed, physically (shaker stress) or osmotically

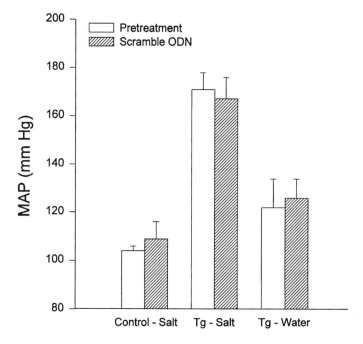

FIGURE 4
PVN injection of scramble ODN had no effect on MAP in control and Tg rats consuming 2% NaCl or water replete Tg rats.

TABLE 2
Characteristics of Cellular Responses
to Antisense Treatment

Basal parameters are unchanged	Stimulated responses are attenuated
Blood pressure	Stress
Heart rate	Osmotic
Fluid intake	Suckling
Peptide secretion	Angiotensin
Electrophysioogy	Neurochemical

(salt intake), were differences noted. The responses were rapid, occurring within hours of administration of antisense, as demonstrated in numerous other studies.[4,6,7,32,47,48] Specific responses were also observed for drinking behavior after treatment with angiotensin AT-1 antisense. Basal intake was not altered, while angiotensin stimulated intake was abolished and cholinergic induced intake unchanged.[29] Examination of electrophysiological properties of supraoptic neurons also verified that peptide antisense treatment alters stimulated, but not basal cellular activity.[46] The

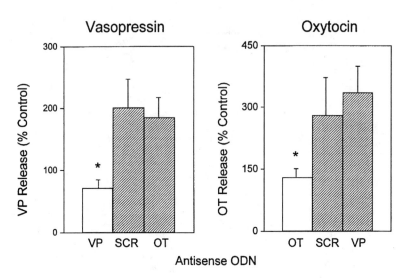

FIGURE 5

Effect of vasopressin (VP), oxytocin (OT) AS or scramble ODN on osmotically induced (NaCl) peptide release in perfused hypothalamic/posterior pituitary explants. There was a specific reduction in peptide release, $p = <0.05$.

effects were specific for antisense and cell type, such that, oxytocin antisense attenuated suckling and cholecystokinin-induced neural activity in putative oxytocin neurons, while vasopressin antisense did not. Peptide and stimulus-specific effects were also observed when hypothalamic/posterior pituitary explants were superfused with oxytocin and vasopressin antisense.[11] Oxytocin antisense attenuated osmotically stimulated oxytocin release, but not vasopressin secretion, with reciprocal results seen after vasopressin antisense treatment (Figure 5). Again, there were no changes in basal secretion. The blockade was not universal for all release stimuli since KCl-induced secretion was not changed in the antisense treated explants. These results discount the idea that antisense exerts a non-specific action on stimulus-secretion coupling, the final step in any secretory response. Furthermore, the fact that there were no changes in pituitary peptide stores or basal secretion suggests that the antisense may have selective actions on a pool of rapidly synthesized or readily releasable peptides. This is consistent with studies which used protein synthesis inhibitors to demonstrate an attenuation of the milk-ejection reflex[49] and depolarization-stimulated catecholamine release.[50]

In conclusion, the mechanism by which exogenously administered antisense ODNs specifically alter cellular responses is not understood. The rapid changes (within hours) and attenuation of neurally mediated stimuli does not easily fit with the classical scheme of long term modulation of intracellular synthesis and secretion by blockade of mRNA transcription and/or translation. A non-antisense mechanism, as yet to be determined, remains a possibility although the sequence and cellular specificity argues against this. It may also be that cellular physiology, including modulation of afferent input, is sensitive to genetic disruption and the consequent changes in neuropeptide/receptor synthesis. However, the signaling pathways which could mediate such a response are a mystery.

Acknowledgments

This work was supported by NIH grant HL43178 (MM), HL 51952 and American Heart Association NC grant 95GS34 (PL). We would also like to acknowledge the scientific and editorial contributions of Drs. Michael F. Callahan, Mike Ludwig, Celia Sladek, Aldo B. Lucion, Debra I. Diz and Ms. Beverly McClellan, Cindy Barrett and Sue Bosch.

References

1. Morris, M. and Lucion, A. B., Antisense oligonucleotides in the study of neuroendocrine systems, *J. Neuroendocrinol.*, 7, 493, 1995.
2. Phillips, M. I. and Gyurko, R., *In vivo* applications of antisense oligonucleotides for peptide research, *Reg. Pep.*, 59, 131, 1995.
3. Landgraf, R., Antisense targeting in behavioral neuroendocrinology, *J. Endocrinol.*, 151, 333, 1996.
4. Neumann, I., Porter, D. W. F., Landgraf, R., and Pittman, Q. J., Rapid effect on suckling of an oxytocin antisense oligonucleotide administered into rat supraoptic nucleus, *Am. J. Physiol.*, 265, R923, 1994.
5. Landgraf, R., Gerstberger, R., Montkowski, A., Probst, J. C., Wotjak, C. T., Holsboer, F., and Engelmann, M., V1 vasopressin receptor antisense oligodeoxynucleotide into septum reduces vasopressin binding, social discrimination abilities, and anxiety-related behavior in rats, *J. Neurosci.*, 15, 4250, 1995.
6. Spampinato, S., Canossa, M., Carboni, L., Campana, G., Leanza, G., and Ferri, S., Inhibition of proopiomelanocortin expression by an oligodeoxynucleotide complementary to β endorphin mRNA, *Proc. Natl. Acad. Sci. USA,* 91, 8072, 1994.
7. Sakai, R. R., Ma, L. Y., He, P. F., and Fluharty, S. J., Intracerebroventricular administration of angiotensin type 1 (AT$_1$) receptor antisense oligonucleotides attenuate thirst in the rat, *Reg. Pep.*, 59, 183, 1995.
8. Heilig, M., Antisense inhibition of neuropeptide Y (NPY)-Y1 receptor expression blocks the anxiolytic-like action of NPY in amygdala and paradoxically increases feeding, *Reg. Pep.*, 59, 201, 1995.
9. Morris, M., Li, P., Callahan, M. F., and Lucion A. B., Central oxytocin mediates stress-induced tachycardia, *J. Neuroendocrinol.*, 7, 455, 1995.
10. Giovenardi, M., Padoin, M. J., Cadore, L. P., and Lucion, A. B., Hypothalamic paraventricular nucleus, oxytocin, and maternal aggression in rats, *NY Acad. Sci.*, 807, 606, 1997.
11. Ludwig, M., Morris, M., and Sladek, C. D., Effects of antisense oligodeoxynucleotides on peptide release from hypothalamoneurohypophysial explants, *Am. J. Physiol.*, 272, R1441, 1997.

12. Li, P., Morris, M., Diz, D. I., Ferrario, C. M., Ganten, D., and Callahan, M. F., Role of paraventricular angiotensin AT1 receptors in salt-sensitive hypertension in mRen-2 transgenic rats, *Am. J. Physiol.*, 270, R1178, 1996.

13. Wahlestedt, C., Pich, E. M., Koob, G. F., Yee, F., and Heilig, M., Modulation of anxiety and neuropeptide Y1 receptors by antisense oligodeoxynucleotides, *Science*, 259, 528, 1993.

14. Zhou, L. W., Zhang, S. P., Qin, Z. H., and Weiss, B., *In vivo* administration of an oligodeoxynucleotide antisense to the D2 dopamine receptor messenger RNA inhibits D2 dopamine receptor-mediated behavior and the expression of D2 dopamine receptors in mouse striatum, *J. Pharmacol. Exper. Therapeut.*, 268, 1015, 1994.

15 Meeker, R. B., LeGrand, G., Ramirez, J., Smith, T., and Shih, Y. H., Antisense vasopressin oligonucleotides: uptake, turnover, distribution, toxicity and behavioral effects, *J. Neuroendocrinol.*, 7, 419, 1995.

16. Yee, F., Ericson, H., Reis, D. J., and Wahlestedt, C., Cellular uptake of intracerebroventricularly administered biotin- or digoxigenin-labeled antisense oligonucleotides in the rat, *Cell. Molec. Neurobiol.*, 14, 475, 1994.

17. Whitesell, L., Geselowitz, D., Chavany, C., Fahmy, B., Walbridge, S., Alger, J. R., and Neckers, L. M., Stability, clearance, and disposition of intraventricularly administered oligodeoxynucleotides: Implications for therapeutic application within the central nervous system, *Proc. Natl. Acad. Sci. USA*, 90, 4665, 1993.

18. Ogawa, S., Brown, H. E., Okano, H. J., and Pfaff, D. W., Cellular uptake of intracerebrally administered oligonucleotides in mouse brain, *Reg. Pep.*, 59, 143, 1995.

19. Yaida, Y. and Nowak, J. T. S., Distribution of phosphodiester and phosphorothioate oligonucleotides in rat brain after intraventricular and intrahippocampal administration determined by *in situ* hybridization, *Reg. Pep.*, 59, 193, 1995.

20. Holopainen, I. and Wojcik, W. J., A specific antisense oligodeoxynucleotide to mRNAs encoding receptors with seven transmembrane spanning regions decreases muscarine m_2 and y-aminobutyric acid$_B$ receptors in rat cerebellar granule cells, *J. Pharmacol. Exper. Therapeut.*, 264, 423, 1992.

21. Stein, C. A., Anti-sense oligodeoxynucleotides — Promises and Pitfalls, *Leukemia*, 967, 1992.

22. Walder, R. and Walder, J., Role of RNaseH in hybrid-arrested translation of antisense oligonucleotides, *Proc. Natl. Acad. Sci. USA*, 85, 5011, 1988.

23. Wahlestedt, C., Golanov, E., Yamamoto, S., Yee, F., Ericson, H., Inturrisi, C. E., and Reis, D. J., Antisense oligodeoxynucleotides to NMDA-R1 receptor channel protect cortical neurons from excitotoxicity and reduce focal ischaemic infarctions, *Nature*, 363, 260, 1993.

24. Liebsch, G., Landgraf, R., Gerstberger, R., Probst, J. C., Wotjak, C. T., Engelmann, M., Holsboer, F., and Montkowski, A., Chronic infusion of a CRH$_1$ receptor antisense oligodeoxynucleotide into the central nucleus of the amygdala reduced anxiety-related behavior in socially defeated rats, *Reg. Pep.*, 59, 229, 1995.

25. Skutella, T., Probst, J. C., Caldwell, J. D., Pederson, C. A., and Jirikowski, G. F., Antisense oligodeoxynucleotide complementary to oxytocin mRNA blocks lactation in rats, *Exper. Clin. Endocrinol.*, 103, 191, 1995.

26. Probst, J. C. and Skutella, T., Elevated messenger RNA levels after antisense oligode-oxynucleotide treatment *in vitro* and *in vivo*, *Biochem. Biophys. Res. Commun.*, 225, 861, 1996.

27. Kalra, P. S., Bonavera, J. J., and Kalra, S. P., Central administration of antisense oli-godeoxynucleotides to neuropeptide Y (NPY) mRNA reveals the critical role of newly synthesized NPY in regulation of LHRH release, *Reg. Pep.*, 59, 215, 1995.

28. McCarthy, M. M., Kleopoulos, S. P., Mobbs, C. V., and Pfaff, D. W., Infusion of anti-sense oligodeoxynucleotides to the oxytocin receptor in the ventromedial hypothalamus reduces estrogen-induced sexual receptivity and oxytocin receptor binding in the female rat, *Neuroendocrinology*, 59, 432, 1994.

29. Sakai, R. R., He, P. F., Yang, X. D., Ma, L. Y., Guo, Y. F., Reilly, J. J., Moga, C. N., and Fluharty, S. J., Intracerebroventricular administration of AT_1 receptor antisense oligonucleotides inhibits the behavioral actions of angiotensin II, *J. Neurochem.*, 62, 2053, 1994.

30. Ji, R. R., Zhang, Q., Bedecs, K., Arvidsson, J., Zhang, X., Xu, X. J., Wiesenfeld Hallin, Z., Bartfai, T., and Hokfelt, T., Galanin antisense oligonucleotides reduce galanin levels in dorsal root ganglia and induce autotomy in rats after axotomy, *Proc. Natl. Acad. Sci. USA,* 91, 12540, 1994.

31. Qin, Z. H., Zhou, L. W., Zhang, S. P., Wang, Y., and Weiss, B., D2 dopamine receptor antisense oligodeoxynucleotide inhibits the synthesis of a functional pool of D2 dopam-ine receptors, *Molec. Pharmacol.*, 48, 730,1995.

32. Gyurko, R., Wielbo, D., and Phillips, M. I., Antisense inhibition of AT1 receptor mRNA and angiotensinogen mRNA in the brain of spontaneously hypertensive rats reduces hypertension of neurogenic origin, *Reg. Pep.*, 49,167, 1993.

33. Berrow, N. S., Campbell, V., Fitzgerald, E. M., Brickley, K., and Dolphin, A. C., Anti-sense depletion of B-subunits modulates the biophysical and pharmacological proper-ties of neuronal calcium channels, *J. Physiol.*, 482, 481, 1995.

34. Ohkura, N., Hijikuro, M., and Miki, K., Antisense oligonucleotide to NOR-1, a novel orphan nuclear receptor, induces migration and neurite extension of cultured forebrain cells, *Molec. Brain Res.*, 35, 309, 1996.

35. Neumann, I., Antisense oligonucleotides in neuroendocrinology: enthusiasm and frus-tration, *Neurochem. Intl.,* 31, 363, 1997.

36. Karle, J., Witt, M., and Nielsen, M., Antisense oligonucleotide to $GABA_A$ receptor y2 subunit induces loss of neurons in rat hippocampus, *Neurosci. Lett.*, 202, 97, 1995.

37. Skutella, T., Stohr, T., Probst, J. C., Ramalho-Ortigao, F. J., Holsboer, F., and Jirikowski, G. F., Antisense oligonucleotides for *in vivo* targeting of corticotropin-releasing hor-mone mRNA: Comparison of phosphorothioate and 3'-inverted probe performance, *Hormone Metabol. Res.*, 26, 460, 1994.

38. Chiasson, B. J., Armstrong, J. N., Hooper, M. L., Murphy, P. R., and Robertson, H. A., The application of antisense oligonucleotide technology to the brain: Some pitfalls, *Cell. Molec. Neurobiol.*, 14, 507, 1994.

39. Perez, J. R., Li, Y., Stein, C. A., Majumder, S., van Oorschot, A., and Narayanan, R., Sequence-independent induction of Sp1 transcription factor activity by phosphorothio-ate oligodeoxynucleotides, *Proc. Natl. Acad. Sci. USA,* 91, 5957, 1994.

40. Guvakova, M. A., Yakubov, L. A., Vlodavsky, I., Tonkinson, J. L., and Stein, C. A., Phosphorothioate oligodeoxynucleotides bind to basic fibroblast growth factor, inhibit its binding to cell surface receptors, and remove it from low affinity binding sites on extracellular matrix, *J. Biol. Chem.*, 270, 2620, 1995.

41. Wang, W., Chen, H. J., Schwartz, A., Cannon, P. J., Stein, C. A., and Rabbani, L. E., Sequence-independent inhibition of *in vitro* vascular smooth muscle cell proliferation, migration, and *in vivo* neointimal formation by phosphorothioate oligodeoxynucle- otides, *J. Clin. Invest.*, 98, 443, 1996.

42. Stein, C. A. and Cheng, Y. C., Antisense oligonucleotides as therapeutic agents — is the bullet really magical?, *Science*, 261, 1004, 1993.

43. Gao, W. Y., Han, F. S., Storm, C., Egan, W., and Cheng, Y. C., Phosphorothioate oligonucleotides are inhibitors of human DNA polymerases and RNase H: implications for antisense technology, *Molec. Pharmacol.*, 41, 223, 1992.

44. Morris, M., Bastos, R., and Antunes-Rodrigues, J., Oxytocin antisense oligonucleotide reduces stress-induced atrial natriuretic peptide secretion, *Hypertension*, 29, 863, 1997.

45. Cirelli, C., Pompeiano, M., Arrighi, P., and Tononi, G., Sleep-waking changes after *c-fos* antisense injections in the medial preoptic area, *Neuroreport*, 6, 801, 1995.

46. Neuman, I., Kremarik, P., and Pittman, Q. J., Acute, sequence-specific effects of oxy- tocin and vasopressin antisense oligonucleotides on neuronal responses, *Neuroscience*, 69, 997, 1995.

47. Meng, H., Wielbo, D., Gyurko, R., and Phillips, M. I., Antisense oligonucleotide to AT1 receptor mRNA inhibits central angiotensin induced thirst and vasopressin, *Reg. Pep.*, 54, 543, 1994.

48. Morris, M., Li, P., Barrett, C., and Callahan, M. F., Oxytocin antisense reduces salt intake in the baroreceptor-denervated rat, *Reg. Pep.*, 59, 261, 1995.

49. Wilson, B. C., Salch, T. M., and Pittman, Q. J., Short-latency effects of protein synthesis inhibition, *Soc. Neurosci. Abstr.*, 22, 627, 1996.

50. Cardenas, A. M., Kuijpers, G. A. J., and Pollard, H. B., Effect of protein synthesis inhibitors on synexin levels and secretory response in bovine adrenal medullary chro- maffin cells, *Biochim. Biophys. Acta*, 1234, 255, 1995.

Chapter 9

The Role of Biogenic Amines in Neuroendocrine Regulation in Conscious Rats

György Bagdy

Contents

0-8493-3363-6/98/$0.00+$.50
© 1998 by CRC Press LLC

145

I. Introduction

Recent advances in the knowledge of neuronal regulation of endocrine function have provided evidence that all biogenic amine neurotransmitters participate in the regulation of the hypothalamo-hypophyseal systems.[1-8] The role of monoamine neurotransmitters in the regulation of the hypothalamic-pituitary-adrenocortical (HPA) axis is among the most extensively studied, but still controversial processes.[1,5,6,9-11] This chapter focuses on the *in vivo*, conscious rat methods used in the studies to examine the influence of biogenic amines on neuroendocrine functions. As an example, a short review of the role of monoamines in the regulation of the HPA axis will be given. This review is not comprehensive, but highlights conflicting data and conclusions. The rest of the chapter deals with physiological and methodological aspects related to the disagreements. Examples for limitations and advantages of conscious rat studies and their comparisons to molecular, cellular, *in vitro* or other *in vivo* techniques are discussed, and neglected details leading to uncertain or false conclusion are emphasized.

II. Monoamine Neurotransmitters Regulating the Hypothalamic-Pituitary-Adrenocortical Axis

The HPA axis plays a key role in the integration of the responses to stress through effects on homeostatic mechanisms. These mechanisms include interactions between the central nervous system, endocrine and immune functions.[6,12-14] Corticotropin-releasing hormone (CRH) and arginine vasopressin (AVP) neurosecretory neurones located in the parvocellular region of the hypothalamic paraventricular nucleus (PVN) are the most important corticotropin-releasing factors. Mechanisms that regulate the activity of the HPA axis through the corticotropin-releasing neurosecretory system include also monoamine neurotransmitters. Despite the voluminous literature regarding the regulation of the HPA axis by monoamine neurotransmitters there are still several unclear, conflicting conclusions. These include the general stimulatory or inhibitory role, the locations and subtypes of catecholamine and serotonin (5-hydroxytryptamine, 5-HT) receptors, and their physiological involvement in basal or different stressor-induced responses of the HPA axis.

A. Norepinephrine and Epinephrine

Cell bodies of the norepinephrine-containing neurones that innervate the parvocellular neurones of the PVN are found in the lower brainstem, namely, the caudal nucleus of the solitary tract (A2 cell group) of the dorsolateral medulla, with some contributions from the A1 cell group and the locus coeruleus (A6).[15,16] Epinephrine-containing neurones originate from cell groups in the rostral ventral medulla (C1) rostral nucleus of the solitary tract (C2) and C3.[2,17] Elevations in catecholamine synthesizing enzymes and increased catecholamine turnover in the paraventricular nucleus during stress has been described.[18] The adenohypophysis does not receive catecholaminergic innervation, but neural and intermediate lobes do.[19]

Several studies have shown a stimulatory role of catecholamines on the activity of the HPA axis.[11,18,20-23] α-1 Receptors are believed to mediate the stimulatory effects of ascending catecholaminergic pathways. The physiological significance of the stimulant α-1 adrenoceptors has been demonstrated in humans.[10] ACTH response to ether stress and the CRF-41 response to hypotension were reduced by intracerebroventricular injection of α-1 receptor antagonists in rats, and norepinephrine-induced CRH release explanted hypothalami was also reduced by phentolamine.[10,24]

Despite the evidence of significant stimulatory role catecholamines mediated by α-1 adrenoceptors several conflicting data are also available. Several studies have described inhibitory actions, and the role of α-2 and β adrenoceptors in the control ACTH secretion is still controversial. Activation of these types of adrenoceptors can result in either stimulation or inhibition of the HPA axis.[6,10] In addition, most studies using reserpine as a catecholamine depletor conclude that catecholamines are inhibitory to the HPA axis.[6,25] The explanation may be that the effect of et least the α-2 adrenoceptors depend very much on the activity of the catecholaminergic input itself, and that reserpine as well as α adrenoceptor antagonist treatments are stressors that activate the HPA axis of the animal.

The possible role of circulating catecholamines, especially epinephrine in the adrenocortical activation during stress was revived in the 1980s. Several groups have confirmed that peripheral injections of β-adrenoceptor agonists stimulate the secretion of ACTH, β-endorphin and other pro-opiomelanocortin-derived peptides from the anterior and the intermediate lobes of the pituitary.[10] In addition, surprisingly, subtype-selective 5-HT agonists simultaneously activate release of catecholamines, mostly epinephrine from the adrenal medulla, and epinephrine and ACTH responses show significant correlation.[4] Further analysis of the mechanisms clearly demonstrated that circulating catecholamines have no significant direct effect on pituitary ACTH release. Thus, parallel central regulatory processes of the HPA axis and the sympathoadrenal system after acute serotonergic or central adrenergic receptor activation are likely, and delayed direct interaction between the two systems is also possible.[4,10,18,26-29] There is, however, no question that in general central catecholamines play a key role in the regulation of HPA axis.

B. Serotonin

Serotonergic projections to the PVN have been reported to originate from the dorsal and median raphe with major contributions of three distinct serotonergic cell groups, the B7, B8 and B9 nuclei.[15-30] Liposits et al.[31] have shown overlapping of 5-HT-containing axons and CRF-immunoreactive neuronal cell bodies in the paraventricular nucleus. The exact origin of these serotonergic terminals are still uncertain.[5]

Early literature on the effect of 5-HT neurotransmission on the function and activity of HPA axis was confusing and contradictory, partly because of the poor knowledge of the 5-HT receptor pharmacology and of the relative nonspecificity of the pharmacologic tools that were then available. Within the last decade, a large number of 5-HT receptors with different anatomical localizations and functions have been identified.[32] 5-HT receptors are now divided into seven major groups (5-HT$_{1-7}$) and 5-HT$_1$, 5-HT$_2$ and 5-HT$_5$ families are further subdivided. It has become clear that CRH/ACTH, β-endorphin, prolactin, vasopressin and oxytocin respones to 5-HT are mediated by different receptor subtypes and by different mechanisms.[5,33-36] Some subtypes of 5-HT$_1$ and 5-HT$_2$ families and 5-HT$_3$ receptors have been extensively studied regarding their role in regulation of HPA axis. In general, 5-HT as well as 5-HT$_{1A}$ and 5-HT$_2$ receptor subtype selective agonists increased CRH release from explanted rat hypothalami *in vitro*, and ACTH and corticosterone secretion *in vivo*.[4,37-38] Pituitary stalk transection, anti-CRH serum, CRH antagonists and lesion of the hypothalamic paraventricular nucleus decreased or blocked ACTH and corticosterone responses, suggesting that mainly CRH mediates the effects of 5-HT$_{1A}$ and 5-HT$_{2A}$ and 5-HT$_{2C}$ receptor stimulation of the HPA axis.[35,39-41] The effect of PVN lesion on 5-HT$_{1A}$ agonist ipsapirone-induced (2.0 mg/kg, i.v.) ACTH and corticosterone responses are shown in Figure 1, the effects of the lesion on 5-HT$_{1A}$ and 5-HT$_{2A}$ and 5-HT$_{2C}$ receptor mediated ACTH and corticosterone responses are summerized in Table 1. 5-HT$_3$ receptors do not seem to participate in 5-HT-mediated CRH release studied in explanted hypothalami *in vitro*,[37] nor in the 5-HT-releaser p-chloroamphetamine-induced ACTH and corticosterone responses *in vivo*.[42] Thus, according to most studies, 5-HT$_{1A}$, 5-HT$_{2A}$ and 5-HT$_{2C}$ receptors stimulate the activity of the HPA axis mainly by a central, CRH dependent mechanism. 5-HT$_3$ receptors do not seem to play an important role in the regulation of the HPA axis.

Regardless the evidence of stimulatory role of 5-HT and activation of some of its receptor subtypes there are several questions and conflicting results. Some studies have shown an inhibitory role of 5-HT$_{1A}$ receptor stimulation on the basal, unstimulated corticosterone concentration or on the response to stress.[43,44] Some disrepancy exists also between the *in vitro* and *in vivo* results. For example, 5-HT$_{1A}$ agonists stimulate ACTH release from cultured pituitary cells *in vitro*, but lesion studies found no or very little ACTH/corticosterone responses independent from the hypothalamus *in vivo*.[35,36,45] Also 5-HT$_2$ receptor agonists stimulated ACTH release that suggests that such drugs might act at the pituitary level as well as on the hypothalamic level.[45] Indeed, 5-HT$_{2A}$ receptor-mediated ACTH and corticosterone responses were attenuated, but not completely blocked by pituitary stalk transection, anti-CRH serum, CRH antagonists and lesion of the hypothalamic paraventricular nucleus, which suggests that other hypothalamic and/or peripheral sites contribute to DOI-induced paraventricular nucleus-mediated activation of the HPA axis.[35,39] In addition to hypothalamic CRH and vasopressin, direct pituitary and/or adrenal effects have been suggested.[35,39,40,46-48] In addition, inhibition of CRH-induced ACTH release by

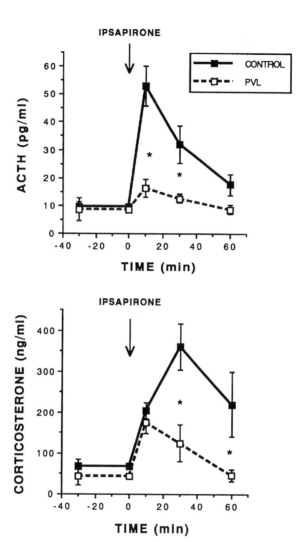

FIGURE 1

Effect of the surgical lesion of the hypothalamic paraventricular nucleus on the 5-HT$_{1A}$ receptor agonist ipsapirone-induced (2.0 mg/kg, i.v.) ACTH and corticosterone responses. Asterisk indicates significant change compared to control, $p < 0.05$. The lesion caused an almost complete blockade of ACTH response, thus the paraventricular nucleus mediates the 5-HT$_{1A}$ receptor agonist-induced responses.

5-HT$_{1A}$ and 5-HT$_{2A}$ receptors and stimulation of vasopressin-induced ACTH release by 5-HT$_{2C}$ receptors have been found *in vitro*.[45] The non-CRH mediated central and peripheral part of the 5-HT2A receptor mediated corticosterone response needs to be clarified, too. Interestingly, 5-HT$_3$ receptor mediated increases in ACTH release from cultured anterior pituitary preparation, and corticosterone responses *in vivo* also have been described.[49,50]

TABLE 1
Effects of Surgical Lesion of the Paraventricular
Nucleus on 5-HT Agonist-Induced Hormone
Responses[a]

5-HT agonist (Receptor subtype)	ACTH	Corticosterone
Ipsapirone (5-HT$_{1A}$)	⇓[b,c]	⇓[b,c,d]
DOI (5-HT$_{2A}$)	—[f]	⇓[c,d]
m-CPP (5-HT$_{2C}$)	⇓[c]	⇓[c,e]

[a] Decrease (⇓) is based on % changes in integrated responses over time, the net
area under the curve in lesioned rats compared to controls.
[b] Data from Section 2.2.
[c] Data from Reference 36.
[d] Data from Reference 35.
[e] Data from Reference 41.
[f] No data.

Modified from *Behav. Brain Res.*, 73, G. Bagdy, Role of the hypothalamic paraventricular nucleus in 5-HT$_{1A}$, 5-HT$_{2A}$ and 5-HT$_{2C}$ receptor-mediated oxytocin, prolactin, and ACTH/corticosterone responses, 277-280, 1996. With kind permission
of Elsevier Science — NL, Sara Burgerharstraat 25, 1055 KV Amsterdam, The
Netherlands.

III. Physiological and Methodological Aspects Related to Conflicting Results

A. Biosynthesis and Secretion of Hormones

3. Timing and Correlation of Immediate Early Genes, Hormone mRNA and Release

The hormonal message and its final action depend on a complex machinery. This
includes the first steps in the genetic ordering of hormonal synthesis, the subsequent
posttranslational processing of the peptide, the release of the hormone, the post-
secretory extracellular transport, the receptor mediation of the hormone and subsequent transduction, inactivation and clearence.[51]

Comparison of the time course of stress-induced ACTH secretion and changes
in the expression of the primary CRH and AVP transcripts with markers of representative transcription factor classes led to the conclusion that the activation-induced
protein synthesis occurs long after the termination of secretory responses, allowing
counter-regulatory and/or other modulatory events.[52] In addition, despite robust
increases in CRH immunostaining and primary CRH and AVP transcripts no reliable
change of CRH mRNA was found in hypophysiotropic neurones after ether stress
or unilateral knife cuts in the lower brainstem.[52-54] Recent data provided evidence
that basal but not immobilization-induced CRH mRNA levels in the PVN are under
tonic stimulatory influence of the lower brainstem.[55] Thus, changes in primary CRH

transcripts, immediate early genes, CRH mRNA, CRH synthesis and secretion may clearly dissociate under different circumstances.

2. Local Control of Secretion

Among several other factors with both inhibitory and stimulatory capacity ample evidence has been accumulated for the existence of local control systems in the anterior pituitary. Paracrine regulation denotes the production and secretion of regulatory substances by one cell and their diffusion to, and activation or possibly inhibition of, a neighboring cell without the intervention of vascular routes. There is evidence of paracrine regulation in corticotroph cells. At least two different populations of corticotrophs exist. The first responds to CRH and the other to AVP.[56] Paracrine mechanism may play an important role in basal ACTH release, in response to CRH and it may contribute to modulatory actions of AVP on corticotrophs that respond to CRH.[56,57]

ACTH is processed and released by the anterior and also by the neurointermediate lobes in the rat.[1] 5-HT is present in the intermediate lobe in at least two different nerve fibers and terminals, in mast cells and in elements in the blood circulating in vessels on the surface of the lobe.[58,59] There is evidence that 5-HT stimulates the release of ACTH and other proopiomelanocortin-derived peptides.[1,47] 5-HT may have a local effect or diffuse to corticotrophs and other cell types located in the close vicinity of the intermediate lobe.

In addition to hypothalamic and pituitary actions 5-HT may directly stimulate corticosterone release. The subtype(s) involved in this effect vary with the species, for example, 5-HT_2 or 5-HT_4 receptors have been found responsible in rat, human and amphibian corticosterone release.[46]

3. Circadian Changes

There is a peak in the activity of the HPA axis at the onset of the normal active period, or wakefulness, and a nadir around the onset of the resting period both in man and the rat.[60] The rise in circadian ACTH and corticosterone concentration is blocked by the administration of anti-CRH serum (see Figure 2).[61] Baseline concentrations of ACTH and corticosterone are, however, not affected by anti-CRH serum as shown on Figure 2.[61] Thus, evening elevations of ACTH and corticosterone depend on release of endogenous CRH, while CRH does not appear to play a significant role in maintaining baseline ACTH and corticosterone levels in the morning in rats.[60-63] That means that in conscious, unstressed rats, resting ACTH and corticosterone concentrations cannot be attenuated by CRH dependent central mechanisms. Neither lesion of the hypothalamic paraventricular nucleus, immunoneutralization of CRH, nor pituitary stalk transection are able to attenuate resting ACTH and corticosterone concentrations in the morning (see Figure 2). Treatment with dexamethasone, that do attenuate basal ACTH and corticosterone concentrations does have a significant action at the level of the pituitary.[61] There is evidence, however, that serotonergic innervation of the suprachiasmatic nucleus is essential for normal circadian variations described above.[64,65]

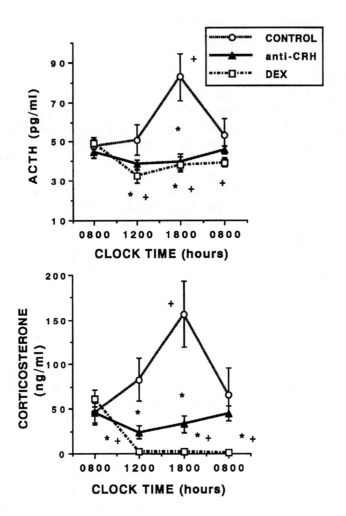

FIGURE 2
Effect of high-titer anti-CRH serum (anti-CRH, 1.0 ml/rat, i.v.) or dexamethasone (DEX, 0.5 mg/kg, i.v.) treatment on diurnal changes in ACTH and corticosterone concentrations in chronically cannulated, freely moving rats. Asterisk indicates significant change compared to control group, $p < 0.05$; plus indicates significant change compared to 08 hr values, $p < 0.05$. Anti-CRH serum failed to cause any significant effect in the morning, but it blocked afternoon elevations in ACTH and corticosterone. (Adapted from Bagdy, G., Chrousos, G. P., and Calogero A. E., *Neuroendocrinology*, 53, 573, 1991, with kind permission of S. Karger AG, P.O.Box, CH-4009 Basel, Switzerland.)

B. Pitfalls of *In Vivo* Methods

1. Quantitative Relationships Between CRH, ACTH and Corticosterone Release

Quantitative relationships of CRH, ACTH and corticosterone responses are extremely important for the interpretation of *in vivo* data. Because the measurement

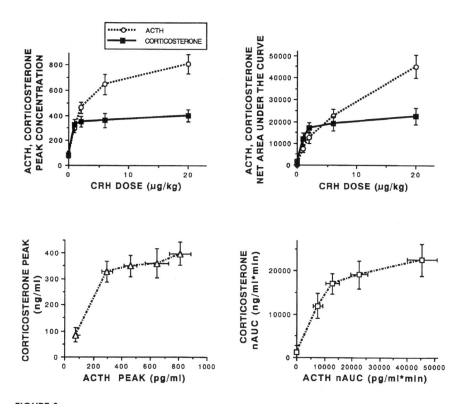

FIGURE 3

Effect of rat CRH (i.v.) administration on plasma ACTH and corticosterone responses in chronically cannulated, freely moving rats. ACTH and corticosterone responses are expressed as peak concentration (left) and net area under the curve (nAUC, calculated from 120 min responses, right). Below: Corticosterone responses are expressed as a function of ACTH responses (left: peak concentrations, right: net area under the curve, nAUC). In general, corticosterone responses are saturable, and because of the early saturation, corticosterone responses do not correlate with moderate or high CRH and ACTH responses.

of the circulating glucocorticoids is relatively easy and inexpensive, several studies use corticosterone as an indicator of the activity of the HPA axis, e.g., coticosterone data serve as a basis for conclusions about the central corticotropin releasing activity. The magnitude of the ACTH response is in general directly related to the intensity of a stress.[12,13] This is, however, not true for corticosterone. Administration of graded doses of rat CRH is a good working model to study the relationship between changes in CRH, ACTH and corticosterone concentration in rats, as shown on Figure 3. While ACTH responses closely resemble a linear correlation in a wide range of CRH, corticosterone responses show a saturation curve. For example, increasing the dose of CRH from 2 to 20 μg/kg caused a 260% increase in ACTH response, while corticosterone response increased by 33% if net area under the curve was used as a measure of the response. Data are even more striking if peaks are compared. Average peak concentration of ACTH increased by about 76%, while corticosterone showed 14%, not a significant increase after 10 times increase of the 2 μg/kg dose of CRH. Thus, in this range, little or no change in corticosterone follows marked changes in CRH. These small changes in corticosterone concentration may be overlapped by the scattering. A similar relationship is evident after the administration of graded doses of different 5-HT agonists to conscious, freely moving rats.[4] Thus, a less than 20%, not significant increase of corticosterone peak could be seen together with a robust 80% increase of ACTH peak with the dose increase from 0.3 to 1.0 mg/kg DOI, i.v.[4] The same effect, although in inverse direction, can be seen after pretreatments. For example, a potent adrenergic or serotergic antagonist that attenuates the ACTH response by more than 75%, might be considered ineffective, based only on the corticosterone data. This fact clearly explains several discrepant ACTH and corticosterone responses. Thus, a lack of significant change in corticosterone response does not demonstrate the lack of change in ACTH, and even less in CRH release.[4,36,39,40]

2. Quantifying the Hormone Responses

Several studies use single point hormone measures instead of time course based on serial blood samples to characterize neuroendocrine responses. In most cases trunk blood is taken by decapitation. This approach has several major pitfalls. Two of them are clearly related to the quantification of the results. First, most of these studies use very few time points after a challenge. This may be acceptable, if time points are chosen at or around the time of the peak hormone concentration. The problem is, however, that pretreatments, lesions, negative feedback and several other factors may change the time of peak concentration (see Figure 1). If this is not carefully studied before the actual challenge, the data may be misleading. Second, quantitative relationships of CRH, ACTH and corticosterone responses are very important as pointed out previously. Even if more than one point, thus, several separate sets of animals are used in the studies, powerful statistics, like analysis of variance for repeated measures or area under the curve cannot be calculated/used. This can be misleading especially if conclusions are based exclusively on corticosterone data. As pointed out in Section 3.2.1., increasing the dose of rat CRH from 1 to 20 µg/kg caused a significant 80% increase in net area under the curve together with a not significant 22% increase in the peak corticosterone concentration (see Figure 3). Thus, changes that are not significant at any separately measured time point may be become significant if area under the curve is calculated. This latter can be calculated only, however, if serial blood samples from the same animals are taken. Calculation based on serial blood sampling has clear advantages.

3. Stress and Negative Feedback

One of the most important physiological regulatory systems of the activity of HPA axis is the glucocorticoid negative feedback. In fact, these feedback effects are very strong, multiphasic and usually quantitatively more prominent then any other physiological inputs.[12,13] Even one single stress response that activates the HPA axis changes the response to a subsequent stress, and multiple or chronic stress situations or elevation of glucocorticoid concentration all result in changes of responsiveness of the HPA axis by several mechanisms. These include the CRH neurosecretory axons, the CRH cell perikarya in the hypothalamic paraventricular nucleus, the limbic system and the circumventricular organ, and several afferent neural inputs.[6,11,12,18,37,66] These actions are mediated by three different receptor subtypes, a high affinity (glucocorticoid type I), a low affinity (glucocorticoid type II), and also by surface receptors.[6] Regarding the onset and duration of the actions at the hypothalamic and pituitary level three different periods or time domains can be distinguished.[12] The complexity of this system can be demonstrated by the interactions of the HPA axis and the brain 5-HT system. 5-HT neurones activate the HPA axis during stress and control circadian elevation. Changes in the concentration of glucocorticoids alter 5-HT turnover in the brain, modify the number of specific 5-HT receptor subtype(s), and neuroendocrine or behavioral responses mediated by them.[27,29]

Some basic, simple but sometimes neglected details regarding the above described mechanisms in *in vivo* studies should be emphasized. First, a single injection or even handling may cause anxiety, pain or discomfort for the animal. Thus, several neural inputs are activated or inhibited. Baseline and stimulated release of hormones are regulated by different mechanisms, e.g., basal, AVP induced and CRH-induced ACTH release is differentially affected by 5-HT$_2$ receptor activation.[45] In addition, the same manipulation causes a stress response, e.g., increase in plasma corticosterone concentration, which may, in turn, cause negative feedback response at different levels. Repeated stress causes several adaptive changes,[6] and even simple repeated injections of saline clearly alter 5-HT-induced release of CRH from explanted rat hypothalami.[67] The other issue is blood sampling. In studies that use decapitation, cages usually have to be removed before the procedure. Even if cages may be left at the previous place, removing the first animal out of the cage activates the rest. Again, baseline vs. stimulated release of hormones are regulated by different mechanisms, therefore, and because of the large scattering, the interpretation of the data may be restricted. For these reasons, the use of chronically cannulated, conscious, freely moving rats in studies of the regulation of the HPA axis has clear advantages, especially when responses to mild or moderate stressors are studied.

4. Chronically Cannulated, Freely Moving Rats: Advantages and Limitations of the Model

Chronically cannulated, freely moving rats can be used in several *in vivo* models, e.g, administration of endogenous substances, circadian studies, drug challenges, brain lesion studies.[4,27,33-36] A brief description of this model is given below. At least two days before the actual challenge, polyethylene cannulae are implanted in the left femoral artery and vein and/or also brain under halothane anesthesia.[4] The animals are housed in separate cages after cannulation. Two days after the surgery animals display physiological circadian changes in HPA axis, and other neuroendocrine and behavioral parameters.[61] Drug administration is possible through the femoral vein. Samples of blood are collected by free flow through the intra-arterial cannula. Blood loss can be compensated by saline, plasma or whole blood replacement.

Compared to other *in vivo* models the use of chronically cannulated, freely moving rats has several advantages. First, the treatment of the animals does not cause any stress. Intravenous catheter allows an i.v. injection or infusion that does not need handling or removing of the animal, and does not cause any pain. It does not cause any arousal reaction or behavioral activation. Second, blood sampling through the arterial catheter does not cause any stress and the animals do not have to be removed. Third, serial blood sampling is possible allowing repeated measure statistics and calculation of area under the curve. Fourth, i.v. administration has several advantages compared to i.p. or s.c. treatment. The availability of the drug is fast, the dose is well quantified, several factors that could impair comparisons (e.g., drug permeation and absorption) are eliminated. Fifth, the number of animals necessary for strong conclusion is a small fraction compared to those studies using decapitation. Also, the plasma obtained by this method is clear, not hemolytic, i.e., more appropriate for analytical techniques. In general, the scattering of the data is also lower using this model. Limitations of this method include the relatively complicated surgery, the time-consuming procedure with about 24 to 48 hr complete recovery, and the large space needed, with standard conditions.[4,61]

IV. Conclusion

In vivo methods are appropriate to study the outcome: the amount and/or actions of the released hormone. They are, however, not relevant to such questions, as how the changes occurred or which steps are involved. Any change in genetic ordering of synthesis, transport, posttranslational processing, or release may ultimately change the amount of released hormone, while alterations in transduction, inactivation, or clearance may modify its action. Furthermore, general physiological regulatory factors, e.g., circadian changes, stress, negative feedback or activation of compensatory mechanisms all have to be taken in account to obtain solid conclusions regarding *in vivo*, cellular or molecular studies.

Acknowledgments

We thank Mrs. E. Anheuer and Mrs. S. Országh for the technical assistance. Part of the data presented here were supported by Hungarian Research Fund Grant T 020500 and Ministry of Welfare Research Grant 03 308/93.

References

1. Tuomisto, J. and Mannisto, P., Neurotransmitter regulation of anterior pituitary hormones, *Pharmacol. Rev.*, 37, 249, 1985.
2. Palkovits, M., Afferents onto neuroendocrine cells, in *Current Topics of Neuroendocrinology*, Ganten, D. and Pfaff, D., Eds., Springer Verlag, Berlin, 1986, 198.

3. Montange, M. and Calas, A., Serotonin and endocrinology, in *Neuronal Serotonin*, Osborne, N. N. and Hamon, M., Eds., Wiley and Sons, Chichester, 1988, 271.

4. Bagdy, G., Calogero, A. E., Murphy, D. L., and Szemeredi, K., Serotonin agonists cause parallel activation of the symphathoadrenomedullary system and the hypothalamo-pituitary-adrenocortical axis in conscious rats, *Endocrinology*, 125, 2664, 1989.

5. Van de Kar, L. D., Neuroendocrine pharmacology of serotonergic (5-HT) neurons, A. *Rev. Pharmac. Toxic.*, 31, 289, 1991.

6. Whitnall, M. H., Regulation of the hypothalamic corticotropin-releasing hormone neu-rosecretory system, *Progr. Neurobiol.*, 40, 573, 1993.

7. Arimura, A., Regulation of growth hormone secretion, in *The Pituitary Gland*, 2nd ed., Imura, H., Ed., Raven Press, New York, 1994, 217.

8. Ben-Jonatan, N., Regulation of prolactin secretion, in *The Pituitary Gland*, 2nd ed., Imura, H., Ed., Raven Press, New York, 1994, 261.

9. Antoni, F. A., Hypothalamic control of adrenocorticotropin secretion: advances since the discovery of 41-residue corticotropin-releasing factor, *Endocrine Rev.*, 7, 351, 1986.

10. Al-Damluji, S., Adrenergic mechanisms in the control of corticotrophin secretion, *J. Endocr.*, 119, 5, 1988.

11. Palkovits, M., Baffi, J. S., and Dvori, S., Neuronal organization of stress response, *Ann. N.Y. Acad. Sci.*, 771, 313, 1995.

12. Keller-Wood, M. E. and Dallman, M. F., Corticosteroid inhibition of ACTH secretion, *Endocrine. Rev.*, 5, 1, 1984.

13. Dallman, M. F., Akana, S. F., Cascio, C. S., Darlington, D. N., Jacobson, L., and Levin, N., Regulation of ACTH secretion: variations on a theme of B, in *Recent Progress in Hormone Research*, Clark, J. H., Ed., Academic Press, New York, 1987, 113.

14. Antoni, F. A., Hypophisiotropic neurones controlling the secretion of corticotropin: is the hypothesis of a final common hypothalamic pathway correct? in *The Control of The Hypothalamo-Pituitary Adrenocortical Axis*, Rose, F. C., Ed., International University Press, Madison, CT, 1989, 317.

15. Palkovits, M., Anatomy of neural pathways affecting CRH secretion, *Ann. N.Y. Acad. Sci.*, 512, 139, 1987.

16. Cunningham, E. T. and Sawchenko, P. E., Anatomical specificity of noradrenergic inputs to the paraventricular and supraoptic nuclei of the rat hypothalamus, *J. Comp. Neurol.*, 274, 60, 1988.

17. Sawchenko, P. E., Swanson, L. W., Grzanna, R., Howe, P. R. C., Bloom, S. R., and Polak, J. M., Colocalization of neuropeptide Y immunoreactivity in brainstem cate-cholaminergic neurons that project to the paraventricular and nucleus of the hypothal-amus, *J. Comp. Neurol.*, 241, 138, 1985.

18. Pacak, K., Palkovits, M., Kvetnansky, R., Yadid, G., Kopin, I. J., and Goldstein, D. S., Effects of various stressors on *in vivo* norepinephrine release in the hypothalamic paraventricular nucleus and on the pituitary-adrenocortical axis, *Ann. N.Y. Acad. Sci.*, 771, 115, 1995.

19. Saavedra, J. M., Palkovits, M., Kizer, J. S., Brownstein, M., and Zivin, J. A., Distribu-tion of biogenic amines and related enzymes in the rat pituitary gland, *J. Neurochem.*, 25, 257, 1975.

20. Krieger, H. P. and Krieger, D. T., Chemical stimulation of the brain: effect on the adrenal corticoid release, *Am. J. Physiol.*, 62, 722, 1970.
21. Tilders, F. J. H., Berkenbosch, F., Vermes, I., Linton, E. A., and Smelik, P. G., Role of epinephrine and vasopressin in the control of the pituitary-adrenal response to stress, *Fed. Proc.*, 44, 155, 1985.
22. Jones, M. T., Gillham, B., Campbell, E. A., Al-Taher, A. R., Chuang, T. T., and DiSciullo, A., Pharmacology of neural pathways affecting CRH secretion, *Ann. N. Y. Acad. Sci.*, 512, 162, 1987.
23. Plotsky, P. M., Regulation of the adrenocortical axis: hypophysiotropic coding, catecholamines and glucocorticoids, in *The Control of the Hypothalamo-Pituitary Adrenocortical Axis*, Rose, F. C., Ed., International University Press, Madison, CT., 1989, 131.
24. Calogero, A. E., Gallucci, W. T., Chrousos, G. P., and Gold, P. W., Catecholamine effects upon rat hypothalamic corticotropin-releasing hormone secretion *in vitro*, *J. Clin. Invest.*, 82, 839, 1988.
25. Ceccatelli, S., Cortes, R., and Hokfelt, T., Effect of reserpine and colchicine on neuropeptide mRNA levels in the rat hypothalamic paraventricular nucleus, *Mol. Brain Res.*, 9, 57, 1991.
26. Szemeredi, K., Bagdy, G., Stull, R., Calogero, A. E., Kopin, I. J., and Goldstein, D. S., Sympathoadrenomedullary inhibition by chronic glucocorticoid treatment in conscious rats, *Endocrinology*, 123, 2585, 1988.
27. Bagdy, G., Calogero, A. E., Aulakh, C. S., Szemeredi, K., and Murphy, D. L., Long-term cortisol treatment impairs behavioral and neuroendocrine responses to 5-HT1 agonists in the rat, *Neuroendocrinology*, 50, 241, 1989.
28. Korte, S. M., Van Duin, S., Bouws, G. A. H., Koolhaas, J. M., and Bohus, B., Involvement of hypothalamic serotonin in activation of the sympathoadrenomedullary system and hypothalamo-pituitary-adrenocortical axis in male Wistar rats, *Eur. J. Pharmacol.*, 197, 225, 1991.
29. Chaouloff, F., Physiopharmacological interactions between stress hormones and central serotonergic systems, *Brain Res. Rev.*, 18, 1, 1993.
30. Sawchenko, P. E., Swanson, L. W., Steinbusch, H. W. M., and Verhofstad, A. A., The distribution and cells of origin of serotonergic inputs to the paraventricular and supraoptic nuclei of the rat, *Brain Res.*, 277, 355, 1983.
31. Liposits, Z. S., Phelix, C., and Paull, W. K., Synaptic interaction of serotonergic axons and corticotropin releasing factor (CRF) synthesizing neurons in the hypothalamic paravenricular nucleus of the rat, A light and electron microscopic immunocytochemical study, *Hystochemistry*, 86, 541, 1987.
32. Hoyer, D., Clarke, D. E., Fozard, J. R., Hartig, P. R., Martin, G. R., Mylecharane, E. J., Saxena, P. R., and Humphrey, P. P. A., VII. International union of pharmacology classification of receptors for 5-hydroxytryptamine (serotonin), *Pharmacol. Rev.*, 46, 157, 1994.
33. Bagdy, G., Sved, A. F., Murphy, D. L., and Szemeredi, K., Pharmacological characterization of serotonin receptor subtypes involved in vasopressin and plasma renin activity responses to serotonin agonists, *Eur. J. Pharmacol.*, 210, 285, 1992.
34. Bagdy, G. and Kalogeras, K. T., Stimulation of 5-HT1A and 5-HT2/5-HT1C receptors induce oxytocin release in the male rat, *Brain Res.*, 611, 330, 1993.

35. Bagdy, G. and Makara, G. B., Hypothalamic paraventricular nucleus lesions differentially affect serotonin-1A (5-HT1A) and 5-HT2 receptor agonist-induced oxytocin, prolactin, and corticosterone responses, *Endocrinology*, 134, 1127, 1994.

36. Bagdy, G., Role of the hypothalamic paraventricular nucleus in 5-HT1A, 5-HT2A and 5-HT2C receptormediated oxytocin, prolactin and ACTH/corticosterone responses, *Behav. Brain Res.*, 73, 277, 1996.

37. Calogero, A. E., Bernardini, R., Margioris, A. N., Bagdy, G., Gallucci, W. T., Munson, P. J., Tamarkin, L., Tomai, T. P., Brady, L., Gold, P. W., and Chrousos, G. P., Effects of serotonergic agonists and antagonists on corticotropin-releasing hormone secretion by explanted rat hypothalami, *Peptides*, 10, 189, 1989.

38. Fuller, R. W., Serotonin receptors and neuroendocrine responses, *Neuropsychopharmacology*, 3, 495, 1990.

39. Calogero, A. E., Bagdy, G., Szemeredi, K., Tartaglia, M. E., Gold, P. W., and Chrousos, G. P., Mechanisms of serotonin receptor agonist-induced activation of the hypothalamic-pituitary-adrenal axis in the rat, *Endocrinology*, 126, 1888, 1990.

40. Rittenhouse, P. A., Bakkum, E. A., Levy, A. D., Li, Q., Carnes, M., and Van de Kar, L. D., Evidence that ACTH secretion is regulated by serotonin 2A/2C (5-HT 2A/2C) receptors, *J. Pharmacol. Exp. Ther.*, 271, 1647, 1994.

41. Bagdy, G. and Makara, G. B., Paraventricular nucleus controls 5-HT2C receptor mediated corticosterone and prolactin but not oxytocin and penile erection responses, *Eur. J. Pharmacol.*, 275, 301, 1995.

42. Levy, A. D., Li, Q., Rittenhouse, P. A., and Van de Kar, L. D., Investigation of the role of 5-HT3 receptors in the secretion of prolactin, ACTH and renin, *Neuroendocrinology*, 58, 65, 1993.

43. Welch, J. E., Farrar, G. E., Dunn, A. J., and Saphier, D., Central 5-HT1A receptors inhibit adrenocortical secretion, *Neuroendocrinology*, 57, 272, 1993.

44. Saphier, D. and Welch, J. E., Effects of the serotonin 1A agonist, 8-hydroxy-2-(di-n-propylamino)tetralin on neurochemical responses to stress, *J. Neurochem.*, 64, 767, 1995.

45. Calogero, A. E., Bagdy, G., Moncada, M. L., and D'Agata, R., Effect of selective serotonin agonists on basal, CRH-, and vasopressin-induced ACTH release *in vitro*, *J. Endocr.*, 136, 381, 1993.

46. Alper, R. H., Evidence for central and peripheral serotonergic control of corticosteronesecretion in the conscious rat, *Neuroendocrinology*, 51, 255, 1990.

47. Bagdy, G., Calogero, A. E., Szemeredi, K., Gomez, M. T., Murphy, D. L., Chrousos, G. P., and Gold, P., Beta-endorphin responses to different serotonin agonists: involvement of corticotropin-releasing hormone, vasopressin and direct pituitary action, *Brain Res.*, 537, 227, 1990.

48. Welch, J. E. and Saphier, D., Central and peripheral mechanisms in the stimulation of adrenocortical secretion by the 5-hydroxytryptamine2 agonist, (+-)-1-(2,5-dimethoxy-4-Iodophenyl)-2-aminopropane, *J. Pharmacol. Exp. Ther.*, 270, 918, 1994.

49. Saphier, D. and Welch, J.E., 5-HT3 receptor activation in the rat increases adrenocortical secretion at the level of the central nervous system, *Neurosci. Res. Comm.*, 14, 167, 1994.

50. Calogero, A. E., Bagdy, G., Burrello, N., Polosa, P., and D'Agata, R., A role for serotonin 3 receptors in the control of adrenocorticotropic hormone release from rat pituitary cell cultures, *Eur. J. Endocrinology*, 133, 251, 1995.

51. Becker, K. L., Nylen, E. S., and Snider, R. H., Endocrinology and the endocrine patient, in *Principles and Practice of Endocrinology and Metabolism*, 2nd ed., Becker, K. L., Ed., J. B. Lippincott, Philadelphia, 1995, chap. 1.

52. Kovacs, K. J. and Sawchenko, P. E., Sequence of stress-induced alterations in indices of synaptic and transcriptional activation in parvocellular neurosecretory neurons, *J. Neurosci.*, 16, 262, 1996.

53. Swanson, L. W. and Simmons, D. M., Differential steroid hormone and neural influences on peptide mRNA levels in CRH cells of the paraventricular nucleus: a hybridization histochemical study in the rat, *J. Comp. Neurol.*, 285, 413, 1989.

54. Watts, A. G., Ether anesthesia differentially affects the content of prepro-corticotropin-releasing hormone,prepro-neurotensin/neuromedin N and prepro-enkephalin mRNAs in the hypothalamic paraventricular nucleus of the rat, *Brain Res.*, 544, 353, 1991.

55. Kiss, A., Palkovits, M., and Aguilera, G., Neural regulation of corticotropin releasing hormone (CRH) and CRH receptor mRNA in the hypothalamic paraventricular nucleus in the rat, Journal of *Neuroendocrinology*, 8, 103, 1996.

56. Denef, C., Paracrine mechanisms in the pituitary, in *The Pituitary Gland*, 2nd ed., Imura, H., Ed., Raven Press, New York, 1994, 351.

57. Schwartz, J., Canny, B., Vale, W. W., and Funder, J. W., Intrapituitary cell-cell communication reglates ACTH secretion, *Neuroendocrinology*, 50, 716, 1989.

58. Palkovits, M., Mezey, E., Chiueh, C. G., Krieger, D. T., Gallatz, K., and Brownstein, M. J., Serotonin-containing elements of the rat pituitary intermediate lobe, *Neuroendocrinology*, 42, 522, 1986.

59. Vanhatolo, S., Soinila, S., Kaartinen, K., and Back, N., Colocalization of dopamine and serotonin in the rat pituitary gland and in the nuclei innervating it, *Brain Res.*, 669, 275, 1995.

60. Krieger, D. T., Rhythms in CRF, ACTH, and corticosteroids, in *Endocrine Rhythms*, Krieger, D. T., Ed., Raven Press, New York, 1979, 123.

61. Bagdy, G., Chrousos, G. P., and Calogero, A. E., Circadian patterns of plasma immunoreactive corticotropin, beta-endorphin, corticosterone and prolactin after immunoneutralization of corticotropin-releasing hormone, *Neuroendocrinology*, 53, 573, 1991.

62. Szafarczyk, A., Hery, M., Laplante, E., Ixart, G., Assenmacher, I., and Kordon, C., Temporal relationship between the circadian rhythmicity in plasma levels of pituitary hormones and in hypothalamic concentrations of releasing factors, *Neuroendocrinology*, 30, 369, 1980.

63. Dallman, M. F., Akana, S. F., Jacobson, L., Levin, N., Cascio, C. S., and Shinsako, J., Characterization of corticosterone feedback regulation of ACTH secretion, *Ann. N. Y. Acad. Sci.*, 512, 402, 1988.

64. Williams, J. H., Miall-Allen, V. M., Klinowski M., and Azmitia, E. C., Effects of microinjections of 5,7-dihydroxytryptamine in the suprachiasmatic nuclei of the rat on serotonin reuptake and the circadian variation of corticosterone levels, *Neuroendocrinology*, 36, 431, 1983.

65. Banky, Z., Molnar, J., Csernus, V., and Halasz, B., Further studies on circadian hormone rhythms after local pharmacological destruction of the serotoninergic innervation of the rat suprachiasmatic region before the onset of the corticosterone rhythm, *Brain Res.*, 445, 222, 1988.

66. Kovacs, K. J. and Makara, G. B., Corticosterone and dexamethasone act at different brain sites to inhibit adrenalectomy-induced adrenocorticotropin hypersecretion, *Brain Res.*, 474, 205, 1988.

67. Calogero, A. E., Liapi, C., and Chrousos, G. P., Hypothalamic and suprahypothalamic effects of prolonged treatment with dexamethasone in the rat, *J. Endocr. Invest.*, 14, 277, 1991.

Chapter **10**

Neuroendocrinology of Stress: Behavioral and Neurobiological Considerations

Béla G. J. Bohus

Contents

I. Introduction

The stress reaction is a common feature of all living beings. Within the animal kingdom including humans, perception of all changes in the external or internal environments elicits a chain of physiological reactions that can be construed as adaptive to the organism. Cannon[1] was the first to study physiological responses to directly threatening environmental stimuli. The "fight or flight" response as natural

0-8493-3363-6/98/$0.00+$.50
© 1998 by CRC Press LLC

reaction to threat is characterized by the activation of the sympatho-adrenal system and the inhibition of the parasympathetic limb of the autonomic nervous system. A sympathetic mass discharge was considered to be responsible for diverse signs of stress such as tachycardia, blood pressure rise, hyperthermia, piloerection, etc. These observations can be considered as roots of stress physiology and pathology. The concept of non-specificity of bodily responses designated as stress to diverse noxious stimuli (stressors) was introduced in the mid nineteen thirties by Hans Selye.[2] Selye's stress theory emphasized the prominence of the adrenal cortical hormone response to stressors. This hormonal response serves to alert the organism to external and internal environmental changes and to preserve homeostasis. Selye's adrenal-oriented concept led stress research to become the domain of (neuro)endocrinologists, a sole domain for the first almost 30 years. The concept was then shared by endocrinologist and psychologists, but a monodisciplinary character was preserved within both disciplines for the next three decades. The emergence of the stress and coping theories in psychology, as developed from the mid nineteen sixties, led to the introduction of the behavioral dimension in the complexity of stress responses.[3] Another emergence in the same period, the foundation of psychoneuroendocrinology, a specialization to amalgamate normal and pathological brain function with the endocrine interieur milieu, resulted in at least a couple of novel developments. First, it was recognized that hormone levels in the circulating blood provide a window into the brain by reflecting emotional and mental states.[4] Studies in laboratory animals, particularly of rodents and primates, and in humans, substantiated this view by showing the involvement of limbic and cortical extra-hypothalamic structures in the regulation of the neuroendocrine system.[5] Second, it emphasized that the brain is the major target of peripheral hormones, and, perhaps more importantly, the nervous system is also a major source of endocrine principles, long regarded as classical hormones.[6] Peptides known for their hormonal properties appeared to be essential transmitters and/or modulators of neural communication. In addition, an array of novel peptides were discovered in the brain and other neural tissues. Despite these developments, stress became an important theoretical framework within psychoneuroendocrinology only during the last one and a half decades.

Surprisingly, many of the developments in psychological and psychoneuroendocrine stress research remained unnoticed for the students of the neuroendocrinology of stress, until rather recently. One of the reasons may be the Selyean tradition; the hypothalamic-pituitary-adrenal (HPA) axis remained long the only focus of interest of fundamental stress research. A select population of neurosecretory neurons of the paraventricular nucleus in the hypothalamus represents the central control of adrenal corticosteroid (CORT) secretion. Importantly, all kinds of external and internal stimuli that are considered as stressors stimulate this neuroendocrine cell population, resulting in the neuronal production and transport of substances via the portal circulation to the corticotrophin cells of the anterior pituitary. Corticotrophin-releasing hormone (CRH) and arginine-vasopressin are the most important secretagogues of the pituitary adrenocorticotrophic hormone (ACTH), which, in turn, drives the adrenal cortex to synthesize and release corticosteroid hormones.[7] Although it was also emphasized that the magnitude of the HPA stress response is markedly controlled by complex emotional, neural and hormonal regulatory mechanisms,[8] the neuroendocrine stress theory was reduced to a straight nonspecific stimulus-response model. The stronger the stressor, the larger is the magnitude of the HPA response, independent of the quality of the stimulus.

An integrated behavioral, physiological and neuroendocrine research has emerged during the last decade which, often using natural stress models like social interactions in animals, suggests both qualitative and quantitative differences in stress response.[9,10] On the other hand, specificity of the stress response emerges from behavioral studies in which the ability to learn about, and to remember the stressful environment plays an important role.[11] These integrated theories, often in contrast with the emergency aspect of the unidimensional neuroendocrine stress views, emphasize the conditioned nature of the stress response, which is a critical aspect in considering stress as a risk factor in mental, neurological, and somatic pathologies in human.[12]

This chapter is devoted to the integrated neuroendocrine response with special reference to the importance of conditioned aspects of the phenomenon. The emphasis is on rodents as experimental models for human stress pathology. Behavioral and neurobiological approaches are used to demonstrate (a) the importance of the selection of methods and stressors in order to interpret the neuroendocrine response; (b) that the primary and conditioned neuroendocrine stress responses share a number of basic properties, but the role of the direct stressor is transferred to subtle cues, and the two responses are served by different neural circuits in the brain.

II. The Neuroendocrine Stress Response

A. Methodological Issues

A discussion of the characteristics of the neuroendocrine stress response should start with a few methodological considerations. It is generally acknowledged that changes in the circulating level of hormones serve as an important objective measure of stress response. It is, however, of crucial importance to know (a) how peripheral blood is collected; (b) the conditions under which the baseline and stress blood samples are withdrawn.

First, in rodents, collecting blood following decapitation or cervical dislocation followed by decapitation has been long the usual method applied. Beside its simplicity, the amount of blood plasma obtained was a crucial point for determining in a specific manner the quantity of circulating hormones. The sensitivity of chemical and radioimmunological techniques has been improved during the last couple of decades: the amount of blood plasma necessary for analysis can no longer be a methodological excuse for using decapitation for stress studies. In addition, the large number of experimental animals necessary to obtain a reasonable number of points for a time curve following decapitation raises both ethical and financial issues.

Development of continuous blood sampling via chronically implanted venous catheter (Reference 13) provides excellent means to obtain even very frequent stress-free blood samples. Retransfusion of blood obtained from non-stressed donor rats may eventually prevent problems arising from the need to withdraw larger volumes of blood in case of multiple hormone determinations. In addition to its ethical and economical justification, stress-free sampling is crucial in obtaining real baseline levels for rapidly released hormones such as the catecholamines (CA) epinephrine (E) and Norepinephrine (NE), and the pituitary hormones ACTH or prolactin (PRL). The advantage of the chronic cannulation technique has its own restriction: the method can nowadays be considered as a routine procedure in rats, but its use in mice is very limited if at all possible. This is an obvious unwanted limitation in future studies on the neuroendocrinology of stress in transgenic animals being almost exclusively mice.

The second issue is the condition under which the baseline and stress samples are collected. The use of chronically implanted cannula allows stress-free baseline blood sampling in home cage conditions in freely moving rats, and monitors minimal changes in the neuroendocrine system. An example for this approach is the use of the defensive burying paradigm in the rat.[14] In this situation a rat is confronted with a stressful event in his/her home cage: an electrifiable rod inserted temporarily through one wall of the cage through which a mild electric shock is given whenever it is touched. The animals can be connected to the externalized pre-implanted cannula before the application of the stressor. This procedure allows the experimenters to obtain a complete base line condition before stress. A territorial aggression paradigm in which the endocrine reaction of a resident rat is measured in response to an intruder[15] also has this advantage of home cage base line measure, similar to the defensive burying paradigm. However, a relatively large home cage is necessary to house the male resident with a female cycling partner (approximately 0.5 m^2) in order to establish territoriality. A recently used seminatural paradigm using a predator (e.g., cat or ferret) as a stressor provides an alternative, biologically relevant social stress procedure (Reference 16). Practically all other paradigms require handling and transfer of animals shortly before the application of stressors: elevation of prestress hormone levels, particularly in the case of rapidly releasable hormones may overshadow the neuroendocrine stress response.

B. Selection of the Stressor: A Behavioral View

A wide array of stimuli has been used to investigate stress mechanisms in animals and in humans. Based on Selye's original hypothesis of the non-specific nature of stress,[2] physical and chemical agents were the first used to put strain on experimental animals. The list of the stimuli included subcutaneous formalin as a pain eliciting stimulus, hemorrhage, cold, insulin-induced hypoglycemia, etc. Some of the stressors such as hemorrhage and insulin could be viewed as stimuli mimicking disturbances of bodily homeostasis. Others such as formalin injection modeled external environmental effects. The emphasis was often on the strength rather than the qualitative properties of the stressor. Restraint elicited by complete immobilization or movement restriction became subsequently a popular experimental stress procedure.

These rather "classical" stressors are still widely used in the present neuroendocrine stress research (Reference 17).

The role of psychological stimuli in eliciting "stress-like" reactions has been sporadically emphasized from the 1950s by a few research groups, particularly in the U.S. and Hungary. The pioneering work of Mirsky, Mason, and Levine in the U.S., and Endroczi and Lissak in Hungary (see Reference 11) showed that emotional stimuli are powerful activators of the pituitary-adrenocortical axis. In terms of multi-hormonal response, Mason[18] was one of the firsts who attempted to conceptualize brain, behavior and endocrine interactions. He suggested that "... the primary mediator underlying the pituitary-adrenal response to diverse stressors ... may simply be the psychological apparatus involved in emotional and arousal reactions to threatening or unpleasant factors in the life situation as a whole". The emotional behavioral view is fully accepted by the investigators of human neuroendocrine stress mechanisms, and the number of animal research papers describing the effects of emotional behavioral stressors and their mechanisms is also steadily increasing. The advantage of the use of behavioral stressors appears to be manifold. First, to measure the behavioral stress response is an additional index of the individuals integrated reactions repertoire. Second, behavioral paradigms permit the study of the relationship between neuroendocrine and other physiological reactions, and the properties of the stressor such as controllability and predictability. Third, behavioral paradigms allow the investigation of individual reaction patterns in relation to coping. Finally, behavioral stress studies allow the selection of experimentally manipulated or species specific natural (social) behaviors as the subject of investigation.

The behavioral stress responses, from the simple fight or flight, or immobility behaviors, to a series of complex spatial and temporal patterns, can be quantified and correlated to other measures of stress reactions. I have suggested that behavioral stress responses may be defined either as situation- and pattern-specific responses or non-specific behavioral reactions.[11] The specific category of behaviors consists of sequences of learned patterns that is adopted by animals including humans as the result of experience.

Some of the behavioral patterns that are emitted in social confrontational situation can also be designated as specific reactions. Aggressive animals use offensive patterns such as chasing, threatening, biting, boxing, keeping down, etc. during social confrontation, whereas the response repertoire of non-aggressive animals consists of fleeing, hiding, freezing, etc. (References 10, 19). The non-specific behavioral responses are independent of the temporal and spatial properties of a certain environment. Displacement behaviors such as grooming, wet-shaking, teeth chattering, defecation, and urination occur in rodents in novel and known environments. Exploration, analgesia, and reflex immobility may also be considered as non-specific responses, but these may habituate or may be conditioned.[11] In sum, the behavioral stress response cannot be viewed as a simple measure of the stressful properties of the environment, but rather as a reflection of the interaction between the environment and its impact on the emotionality and memory of an individual.

How can we translate the behavioral stress profile into the neuroendocrine response pattern? We have formerly argued that in different stressful situations an animal can avoid threat by attempting to cope actively, i.e. by actions such as fight or flight; or passively, i.e. by refraining from action with immobility or submission. Genetic factors in rodents determine very prominently the use of only active or only passive coping strategies, independent of the given social or non-social environment.[9,19] For average populations the characteristics of the environment are the decisive factors of choice between the two strategies. The selected behavioral choice is then reflected in the endocrine reactions.[20] This conclusion was reached in a study using the defensive burying paradigm. On the first day of the test, two groups of rats were allowed to bury the electrified probe by saw-dust bedding, whereas the third group was prevented from burying this rod by removing the bedding material from the home cage. The latter group of rats avoided the rod and showed mostly immobility (freezing) behavior. On the second day, the rats were confronted again with the non-electrified probe under different bedding conditions, and their behavior and endocrine responses were determined. In rats trained and tested under burying conditions — i.e. active coping — burying behavior was accompanied by high plasma NE but low E and CORT levels. The dissociation between NE and E responses suggests selective peripheral autonomic activation that can be ascribed to the association with activity, involving skeletal muscular exertion.[21] In conjunction with passive coping — i.e. rats trained and tested without saw-dust — immobility was accompanied by pituitary-adrenocortical activation resulting in high CORT, but moderate NE and E levels. This suggests a ruling HPA system reaction that is usual in most environments requiring passive coping.[9] The third groups, i.e. rats trained with burying and tested while preventing burying, the stress reaction was the most remarkable: immobility behavior was accompanied by the highest plasma E increase, associated with moderate NE elevation and the highest CORT levels. These animals actually combined characteristics of both types of endocrine response patterns, i.e. high CORT and NE, superimposed by anxiety component (high E) caused by the unpredictable (burying) response prevention.[22]

In sum, relatively subtle environmental modifications provide us with an excellent tool to investigate stress mechanisms *per se,* and the relation to coping or failure to cope with the challenging conditions. However, one should realize that an exclusive behavioral view may lead investigators to neglect the results of a large number of studies with non-behavioral stressors. A behavioral view absolutely favors Mason's view on the necessity of emotional changes to induce the stress reaction.[18] Emotional changes due to immobilization, fasting, heat, exercise, etc. are obvious, but, particularly in experimental animals, it is not always possible to quantify these emotional changes. In other cases such as insulin-induced hypoglycemia, hemorrhage, etc. the emotional tension caused by the stressor is almost impossible to determine. These arguments again open the discussion on the specificity or nonspecificity of stressor and stress response mechanisms. This discussion was ongoing between Mason and Selye in 1975. In Selye's view "all this is explained by conditioning factors and by the fact that not all stressors reach the headquarters ... in the hypothalamus through the same pathways" (cited by Reference 23). Both the aspects of conditioning and functional architecture of the stress responses will be considered later in this chapter.

C. The Primary Neuroendocrine Stress Response

The primary or acute neuroendocrine stress response can simply be described as the organism's first reaction to mostly unknown or unexpected challenge. This corresponds to Selye's alarm phase of the stress. The alarm phase is then followed by a recuperation of the bodily functions provided that the stressor is terminated. Behavioral and physical changes characterize the adaptive stress response.[24] The behavioral component consists of increased arousal and alertness, heightened attention, and suppression of other behaviors such as food intake, reproduction, etc. The physical component is described as redirection of the energy within the stressed body to the central nervous system. In this view, the key elements of the stress system are the hypothalamic-pituitary-adrenal and the sympatho-adrenal systems as controlled by central nervous mechanisms.[25] Certain factors can be recruited in special cases such as inflammation: cytokines or lipid mediators potentiate the activity of the HPA system (e.g., Reference 26). The proper functioning of these two components underlie physiological adaptive changes serving metabolism, gastrointestinal, reproductive, immune, and other bodily functions. This traditional view considers the HPA axis hormones and CAs as the stress hormones responsible for physical and mental health. Disturbances in these neuroendocrine systems entail pathology.

Mason's[18] proposition on a multi-hormonal stress response pattern emphasizing a specific or selective organization of the neuroendocrine stress response seems to be in conflict with a view on a bihormonal stress response. Can hormonal systems other than the HPA axis and sympatho-adrenal (S-A) system be designated as "stress hormones"? Two definitions are proposed. One definition is based on the release characteristics of the hormones. "Stress hormones" are those endocrine factors which are released into the blood circulation in response to divers stressors. The alternative definition considers the functional efficacy of hormones either in adaptive or maladaptive terms. "Stress hormones" in adaptive terms maintain homeostasis by assuring proper conditions to respond and to recover adequately in conjunction with challenge. Maladaptation refers to the tissue damaging property of stress hormones in Selyean terms. The main focus in homeostatic terms is on the metabolic function of various hormones released during stress. If one scrutinizes the divers neuroendocrine stress theories it appears that the aspects of release and functional efficacy remain often un-separated. This is principally correct, unless it is used as a surrogate to explain missing chains in stress mechanisms. This latter is, however, often true because a number of hormones released by stressors do not have clear adaptive or maladaptive effects in strict metabolic terms. However, release and functional efficacy are ununseparated aspects if a hormonal state of the brain is considered from a behavioral perspective as the major target of "stress hormones".[9]

The "stress hormones" of the "release" definition can be divided on a more or less historical basis into two generations.[11] The hypothalamic-pituitary-adrenal axis, (i.e. CRH, ACTH, and CORT) and sympatho-adrenal hormones (i.e. E and NE) belong to the first generation. The second generation is represented by long-known hormones with rather recently discovered stress related release, and by novel hormones or cytokines with stress-induced release properties. Vasopressin, oxytocin, prolactin, gonadotropins, angiotensin II, Neuropeptide Y (NPY), endorphins and enkephalins, neuroactive steroids, and cytokines should be considered.

The stress hormone character of the first generation does not require a long discussion. Two aspects deserve attention. First, whereas circulating corticosteroids were long considered as the only measure of stress-induced HPA activation, both from mechanistic and functional points of view, alterations in circulating CRH and ACTH levels should also be considered due to differential regulation of the different levels of the system.[27] Second, both the magnitude and the quality of the stressors (type of stressor, single or repeated exposure within a session, etc.) should be considered.[22,23,28]

The characteristics of the stress hormone release of the second generation will be discussed only in short. A number of recent excellent papers described the properties of stress-induced release of these hormones. The hypothalamic neuro-secretory nonapeptides vasopressin and oxytocin share a number of structural and organizational properties, but the response to different stressors is not comparable. Oxytocin is released by a large number of stressors that also affect HPA and S-A system activation, whereas vasopressin response is specific for the stressor.[29] Prolactin release is one of the most typical responses to a number of stressors,[30] but surprisingly this pituitary hormone does not receive sufficient attention in stress research. Gonadotropin release in relation to stress appears to be mostly suppressed.[31] However, studies in social-aggressive stress constrains show that, dyadic aggressive interactions — that are undoubtedly of stressful character for both rodent partners (as indicated by HPA and S-A activation) — result in differential changes in pituitary-gonadal axis as shown by plasma testosterone levels. Testosterone increases in the future winner of the fight from the onset of the confrontation, whereas in the to be defeated rat, testosterone level is not changed during the fight, but declines following defeat.[32] Angiotensin II (AII) also can be considered as a new generation of stress hormones because of its rather recently discovered release following physical and psychological stressors. The octapeptide AII is the product of the multi-component renin-angiotensin system.

The aspartyl protease renin is generated in the juxtaglomeluar cells of the kidney. The enzyme cleaves the inactive decapeptide AI from the angiotensinogen precursor molecule. In turn, AI is converted to active AII by angiotensin converting enzyme. That renin release to certain stressors can be blocked by the beta-adrenoceptor blocker propranolol suggest that the release of this hormone is regulated by the sympathetic nervous system. Therefore, it remains to be answered as to whether the release of AII provides additional information to the CA levels about the initial neuroendocrine response.[33] Another new stress hormone of peptidergic nature is NPY. It is produced in sympathetic nerve endings and co-released with catecholamines. It is released during many stress conditions, but one may also question whether NPY release provides extra information. In humans, there is a clear correlation between CA and NPY release, but in rats, immobilization stress (a strong releaser of CA) is not followed by NPY release.[34] The endogenous opioids, like beta-endorphin are usually co-released with other opiomelanocortin fragments

(ACTH, alpha-MSH) in response to various stressors. The level of the shorter opioid fragment met-enkephalin also is increased in the circulating blood of stressed rodents.[35] It is therefore conceivable that endorphins and enkephalins are also stress hormones. Surprisingly, not much attention is focused on these hormones in stress research, except for their role in stress-induced analgesia.[36] A significant discovery of the last two decades is that beside the corticosteroid hormones, other pregnane neuroactive steroids are produced by the adrenal cortex. That diverse stressors increase the level of the neuroactive steroids allopregnenolone, progesterone, and tetrahydro-deoxycorticosterone in the circulation puts these hormonal factors in the category of stress hormones.[37] Finally, the recognition of the role of cytokines as activators of the neuroendocrine stress systems opens new avenues in stress research.[26,38] Interleukins and other cytokines represent the basic cell to cell communication within the immune system. The cytokine interleukin-1 (IL-1) is one of the key modulators of immunological responses to stress, infection and antigenic challenge.[39] IL-1 has potent neuroendocrine actions via peripheral and/or central mechanisms including the HPA axis.[40,41]

In terms of functional efficacy, most 'stress hormones' should be essential regulators of adaptive bodily processes during challenge. Most hormones designated as stress hormones according to the release criterion have receptors in one or more peripheral organ systems such as the cardiovascular, neuromuscular, immune, gastrointestinal, metabolic and energy supplying, systems. The effects of stress hormones can mostly considered anabolic in adaptive terms. Catabolic actions as maladaptive processes are often seen is chronic stress situations.

More importantly, all stress hormones have receptors in the brain, either in neuronal or glial cells. A large number of behavioral and neurobiological data support the view that the brain is a main target of stress hormones.[9,42] The role of stress hormones in behavioral stress reactions may be acute (i.e. during the stress that releases the hormones) or delayed. For example, it has long been known that removal of the pituitary gland is followed by deficits in learning a specific stress response such as avoidance. The deficits can be corrected by the administration of various pituitary stress hormones such as ACTH, alpha-MSH, beta-endorphin, but also by a supplementary therapy with a mixture of peripheral hormones such as corticosterone, thyroxin, and testosterone.[43] A delayed kind of action of stress hormones refers to their role in the immediate post-stress period. In the case of specific behaviors, this period is designated as consolidation of memory of stressful events. Modulation of memory storage by most of the stress hormones has been demonstrated.[44,45,46] For example, the release of both adreno-medullary and adrenal cortical hormones during stress and their action on memory consolidation processes are essential to remember stressful experience.[47] Substantial knowledge has been collected on physiological, cellular, and molecular neurobiological processes underlying the influence of stress and stress hormones on memory processes.[48,49] The process of storage is the basic event leading to short- or long-term remembrance of the stressful event whenever the basic properties of the stressful events, such as cues or contexts reappear. Successful storage means that the event is conditioned.

D. The Conditioned Stress Response

Learning about the nature of social and physical environment, and remembering its characteristics is a basic adaptive capacity of animals, including humans. Learning and memory in adaptive terms assume conditioning processes. Stress as the reaction to environmental changes is an essential component of adaptive alterations and therefore should be conditioned. In turn, the complexity of the stress response assumes that neuroendocrine components of the stress should also be conditioned. This argument seems to be logical, and biologically very essential. It is therefore surprising that the conditioned stress response has not been widely emphasized. Based on non-specificity, conditioning of stress has been frequently interpreted in negative terms. For example, conditioning of immune responses has been interpreted as a highly specific process from which the "nonspecified" stress component should be eliminated.[50] It is therefore a basic question as to whether stress as a complex process is conditioned or, alternatively, whether each of its components, including the neuroendocrine changes are separately learned and retrieved. Unfortunately, this question cannot be easily answered. First, to the authors' best knowledge, this question has not been directly addressed experimentally. Second, this question cannot be viewed independently of the problem of uniformity vs. complexity, i.e. the existence of more than one form of the stress response.[23] Specificity of stress is often interpreted in quantitative terms: the stronger the stressor, the less specific is the stress response. A large number of data based on neurobiological and psychopharmacological approaches has shown that the behavioral, neuroendocrine and physiological (e.g., cardiovascular, temperature, etc.) responses can be separated, i.e. one can be expressed in the absence of the other. However, the findings do not exclude the possibility that the aversiveness of the stressor is the entity that is primarily conditioned, and the different responses are separately organized via the output channels. In order to avoid a speculative discussion, the subject will be discussed in terms of the appearance of the response in relation to the kind of hormone and the kind of conditioning process.

The name of the Russian Ivan Petrovich Pavlov is inevitably connected to the subject of conditioning. Although his work is primarily known for conditioning with food reward, he also emphasized the ability to condition with aversive events, designated as defense conditioning.[51] Today, Pavlovian defense conditioning is usually referred to as fear conditioning.[52] After some early Russian studies in the nineteen sixties, Ader[53] showed the conditioning of CORT responses to illness-induced aversion associated with the sweet taste of saccharin. Conditioned hormone responses were demonstrated in food and water intake paradigms,[54] and in relation to sexual behavior.[55]

As mentioned above, the major recent trend is the use of fear conditioning paradigms to investigate conditioning processes.[52] The basis of this procedure is to pair a neutral stimulus of tone or light with a painful electric footshock. Following several repetitions, the tone or light becomes the conditioned stimulus (CS) to elicit fear of the aversive experience without the actual presentation of the unconditioned painful stimulus. Whereas the behavioral and physiological consequence of fear conditioning is extremely well documented,[52] knowledge on neuroendocrine alterations is very limited.

Our approach to investigating conditioned stress processes is the use of biologically relevant conditions. In addition, coping with the environmental challenge, and the style of coping is an important aspect of our research (see earlier in this chapter). Whereas fear conditioning in the Pavlovian way uses defined cues (tone, light) as conditioned stimuli, mostly in a repetitive fashion, we adapted procedures where the complexity of the physical or social environment serves as a conditioned stimulus. This procedure, usually referred to as contextual conditioning, consists of a single pairing of a given environment with the aversive event. In the context of a non-social environment, the rats receive a single unavoidable footshock in a dark box. The rats are then re-exposed to the same box a day later, but without applying the physical stressor. Contextual conditioning results in freezing as the main behavioral response. In the social context, the experimental rat is confronted with a trained aggressive dominant rat in a neutral environment. The confrontation results, by definition, in the defeat of the experimental rat within the 10-min observation period. A day later the experimental rat is re-exposed to the same environment, but the dominant rat is caged in a small wire mesh cage in order to avoid bodily contact with the dominant animal. The behavioral response of the defeated rat is to avoid the cage of the dominant rat.

The principal features of the neuroendocrine responses in the contextual conditioning are marked rises in plasma E, NE, CORT and PRL levels.[20,56,57] In the social situation, the most marked rise occurs in plasma NE levels with somewhat moderate increase in E and CORT levels.[15]

A fear conditioning procedure similar to our non-social conditions is used by Van de Kar and Gray and associates.[58,59] A repeated exposure to inescapable footshock is applied to rats. This kind of contextual conditioning results in conditioned rise in plasma ACTH, CORT, renin and PRL levels.

Although multiple control groups are used in these experiments, one cannot exclude the possibility that the conditioned neuroendocrine response is somewhat overshadowed by the effect of handling and transfer of the rats to the experimental environment. Experiments in the defensive burying paradigm,[20,22] as described earlier in this chapter, exclude this disadvantage. This conditioning can also be considered as a contextual procedure in which the insertion of the rod is the source of the unconditioned stimulus. In addition, as discussed earlier, the latter paradigm allows the investigator to distinguish between active and passive, coping-related conditioned neuroendocrine responses. Briefly, conditioned burying behavior is accompanied by a marked NE response, whereas freezing-dominated conditioned behavior is characterized by high CORT level.[20] Accordingly, a single hormonal measure in a free-coping situation does not necessarily reflect the conditioned neuroendocrine response.

In sum, conditioning of the neuroendocrine stress response is clearly demonstrated by these experiments. It is remarkable, however, that the list of stress hormones that was studied in conditioning experiments is much shorter than the list of hormones determined for the primary stress response. Recent studies by Onaka et al.[60] provide evidence that oxytocin is released as a consequence of fear conditioning. However, their observation indicating a suppression of vasopressin release is somewhat surprising, and needs further explanation.

III. The Functional Architecture of the Neuroendocrine Stress Response

Extensive research focuses on the neurobiology of stress and related behavioral, physiological, and neuroendocrine events. Actually, the subject deserves a separate review. However, it seems worthwhile to summarize some of the main findings indicating fundamental differences in the functional architecture of primary and conditioned neuroendocrine responses. The discussion is restricted to studies that used behavioral rather then conventional stressors. The main emphasis of these studies is on the role of the limbic system, and particularly of the amygdaloid complex in neuroendocrine stress processes. A number of extensive reviews are available about the neuroanatomical and neurochemical basis of the amygdaloid regulation of neuroendocrine system (e.g. References 61 and 62). Therefore, the discussion in this section will be focused on the functional differences, and will not go into the details on the neural output systems subserving the regulation of release of different hormones.

Studies by Van de Kar et al.[58] showed that destruction of the central nucleus of the amygdala (CEA) by ibotenic acid injection inhibits the elevation of CORT secretion by immobilization (primary stressor) and by contextual conditioning. These lesions also inhibited renin release due to contextual conditioning, whereas the response to immobilization remains unchanged. In contrast, lesion of the lateral amygdaloid nucleus potentiate the renin response to immobilization, whereas conditioned renin response and both primary and conditioned CORT elevations are unaffected. In a subsequent study Gray et al.[59] demonstrated that chemical (ibotenic acid) lesions in the bed nucleus of the stria terminalis (BNST) inhibit the ACTH, CORT and PRL responses to contextual fear conditioning. The same lesion failed to affect plasma renin responses either to fear conditioning or immobilization stress, and the primary ACTH, CORT, and PRL responses to immobilization remained also unchanged. The BNST is intimately connected to the CEA. It is therefore likely that the nerve pathways controlling the pituitary release of ACTH and PRL in a conditioned situation use the hypothalamic output from the BNST.

Our own approach to this question, as reviewed by earlier studies,[63] consisted of electrolytic destruction of the CEA, and the responses were studied in behavioral situations. Bilateral CEA lesions abolish plasma E, NE, CORT and PRL responses to the primary stress of inescapable footshock. All hormones show an initial increase during stress, but the magnitude of the rise is less pronounced, and the post-stress decline to base-line is more rapid. The findings were interpreted to suggest that the CEA is not involved in the initiation, but rather in the amplification and maintenance of the neuroendocrine primary stress response.

The influence of CEA lesions in the conditioned contextual fear paradigm depends on the time of the lesion — i.e. before the learning or after conditioning of the stress response. A CEA lesion produced before learning completely blocks the appearance of the conditioned rise in the plasma CORT and PRL level, whereas the conditioned elevation in the plasma E remained unchanged. In sum, the components of the unconditioned neuroendocrine response (CORT and PRL) are essential to condition the stress reaction, whereas the sympatho-adrenal component of the conditioned response is independent of the functioning of the CEA.

A totally different picture emerges if the rats have acquired the stress response with an intact amygdala, and the CEA is destroyed in the time between the learning and conditioned test. Such a lesion has no effect on the appearance of the conditioned CORT, PRL and E responses. This finding suggests that following consolidation of the stressful experience, a different network is used to express the conditioned neuroendocrine response. The precise architecture of such a network is not known yet. The BNST and frontal, entorhinal, and orbital cortical areas are candidates of a conditioned neuroendocrine circuit. A role for the hippocampus also remains to be shown. This structure has long been known to be involved in both neuroendocrine regulation and contextual conditioning.[5,52] It should be noted that the CEA is important for other components of the conditioned stress response: conditioned physiological (e.g., lowering of the heart rate) and endocrine (e.g., feeding related conditioned insulin release) responses — i.e. changes mediated via the vagus nerve — and freezing behavior are blocked by postlearning damage of the CEA.

In sum, much has been learned about the role of the CEA in primary and conditioned neuroendocrine responses. This structure has plastic properties,[52] but it has not yet been clearly shown that the CEA is the only site where the association between the CS and unconditioned stimulus, leading to the neuroendocrine response takes place. Further knowledge is imperative to understand which structures are common and which are different in the architecture of stress-related neuroendocrine, behavioral and physiological responses.

IV. Concluding Remarks

The conditioned neuroendocrine stress response patterns share a number of similarities with the primary hormonal responses. However, the aversive character of physically damaging stressors is transferred to subtle compound sensory cues or contextual stimuli signaling danger. This transfer is then the main factor that differentiate between the mechanisms underlying the primary and the conditioned response. It is likely that as the result of the conditioning, an amplification occur within the neurobiological networks controlling hormone release. One should clearly separate this amplification from the changes that occur during repeated stress such as immobilization or sound stress. In these cases, there is practically no behavioral coping component, and the response magnitude is mostly habituated to the stimulus.[28] From the psychopathological and stress-pathological viewpoint, conditioning is a more obvious risk than a repetitive occurrence of the same stressor.

An increasing number of observations is available regarding the neuroendocrinology of stress in long-term social context. Experimental or natural colony aggregation in rodents and primates (e.g., References 10 and 64) results in state changes in the neuroendocrine systems, depending upon the social status of the individuals within the group hierarchy. Phasic responses to social stimuli within a group context are probably comparable to conditioned endocrine responses. However, such studies are extremely difficult due to the necessity of frequent and well-timed blood withdrawal. Implantable radiotelemetric recording of physiological parameters such as heart rate, blood pressure, temperature, are already available for such studies. Neuroendocrinology may gain access to this technique by the development of implantable chemosensors for hormone determination.

Finally, there is an increasing number of observations indicating that a single (traumatic) stress leads to long-term alterations, both in the state and the reactivity of the neuroendocrine and physiological systems.[65,66] The increased reactivity may have elements of conditioning, but it is more likely that instead of specific conditioning, the traumatic experience leads to generalization to all kind of stressors. These single stressor paradigms are frequently used to model depression. However, post-traumatic stress disorder (PTSD) is also characterized by an inextinguishable conditioning and generalization component. One should not forget that the activity of the HPA axis may differentiate between the two conditions. In humans, depression is accompanied by a hyperactive HPA system, whereas PTSD is characterized by a hypoactive HPA axis.

In sum, future research should further specify the role of conditioning of the neuroendocrine response in the mechanisms of diverse psychopathological disorders, in order to understand neurobiological mechanisms underlying stress as a risk factor, and to develop more rational pharmacotherapy for stress-related disorders.

References

1. Cannon, W. B., *Bodily Changes in Pain, Hunger, Fear and Rage,* Appleton, New York, 1915.
2. Selye, H., *Stress: The Physiology and Pathology of Exposure to Stress,* Acta, Montreal, 1950.
3. Lazarus, R., *Psychological Stress and the Coping Process,* McGraw-Hill, New York, 1966.
4. Mason, J. W., A review of psychoendocrine research on the pituitary-adrenal cortical system, *Psychosom. Med.,* 30, 567, 1968.
5. Bohus, B., The hippocampus and the pituitary-adrenal system hormones, in *The Hippocampus, Vol. 1, Structure and Function,* Isaacson, R., L. and Pribram, K. H., Eds., Plenum Press, New York, 323, 1975.
6. Bloom, F. E., Neuropeptides, *Sci. Am.,* 245, 148, 1981.
7. Whitehall, M.H. Regulation of the hypothalamic corticotrophin-releasing hormone neurosecretory system. *Progr. Neurobiol.,* 40, 573, 1993.
8. Keller-Wood, M. E. and Dallman, M. F., Corticosteroid inhibition of ACTH secretion, *Endocr. Rev.,* 5, 1, 1984.

9. Bohus, B., Benus, R. F., Fokkema, D. S., Koolhaas, J. M., Nyakas, C., Van Oortmeerssen, G., Prins, A. J. A., De Ruiter, A. J. H., Scheurink, A. J. W., and Steffens, A. B., Neuroendocrine states and behavioral and physiological stress responses, *Progr. Brain Res.*, 69, 445, 1987.

10. Blanchard, D. C., Spencer, R. L., Weiss, S. M., Blanchard, R. J., McEwen, B. S., and Sakai, R. R., Visible burrow system as a model of chronic social stress: behavioral and neuroendocrine correlates, *Psychoneuroendocr.*, 20, 117, 1995.

11. Bohus, B., Neuroendocrine interactions with brain and behavior: a model for psychoneuroimmunology?, in *Breakdown in Human Adaptation, Vol. 2,* Ballieux, R. E., Ed., Martinus Nijhoff, Boston, 638, 1984.

12. Chrousos, G. P. and Gold, P. W., The concepts of stress system disorders: overview of behavioral and physical homeostasis, *J. Am. Med. Ass.*, 267, 1244, 1992.

13. Steffens, A. B., A method for frequent sampling of blood and continuos infusion of fluids in the rat without disturbing the animal, *Physiol. Behav.*, 4, 883, 1969.

14. Treit, D., Pinel, J. P. J., and Terlecki, L. J., Shock intensity and conditioned defensive burying in rats, *Bull. Psychon. Soc.*, 16, 5, 1980.

15. Korte, S. M., Smit, J., Bouws, G. A. H., Koolhaas, J. M., and Bohus, B., Behavioural and neuroendocrine response to psychosocial stress in male rats: the effects of the 5-HT1A agonist ipsapirone, *Horm. Behav.*, 24, 554, 1990.

16. Adamec, R. E. and Shallow, T., Lasting effects on rodent anxiety of a single exposure to a cat, *Physiol. Behav.*, 101, 1993.

17. Kvetnansky, R., Pacak, K., Fukuhara, K., Viskupic, E., Hiremagalur, B., Nankova, B., Goldstein, B. S., Sabban, E. L., and Kopin, I. J., Sympathoadrenal system in stress, *Ann. N. Y. Acad. Sci.*, 771, 131, 1995.

18. Mason, J. W., Historical view of the stress field, *J. Human Stress,* 1, 6, 1975.

19. Benus, R. F., Bohus, B., Koolhaas, J. M., and Van Oortmerssen, G. A., Heritable variation for aggression as a reflection of individual coping strategies, *Experientia,* 47, 1008, 1991.

20. Korte, S. M., Bouws, G. A. H., Koolhaas, J. M., and Bohus, B., Neuroendocrine and behavioural responses during conditioned active and passive behaviour in the defensive burying/probe avoidance paradigm: effects of ipsapirone, *Physiol. Beh.*, 52, 335, 1992.

21. Scheurink, A. J. W. and Steffens, A. B., Central and peripheral control of sympathoadrenal activity and energy metabolism in rats, *Physiol. Behav.*, 48, 909, 1990.

22. De Boer, S. F., Slangen, J. L., and Van Der Gugten, J., Plasma catecholamine and corticosterone levels during active and passive shock-prod avoidance behavior in rats: effects of chlordiazepoxide, *Physiol. Behav.*, 47, 1990.

23. Kopin, I. J., Definitions of stress and sympathetic neuronal responses, *Ann. N. Y. Acad. Sci.*, 771, 19, 1995.

24. Dom, L. D. and Chrousos, G. P., The endocrinology of stress and stress system disorders in adolescence, *Endocrinol. Metab. Clin. N. Am.*, 22, 685, 1993.

25. Stratakis, C. A. and Chrousos, G. P., Neuroendocrinology and pathophysiology of the stress system, *Ann. N. Y. Acad. Sci.*, 771, 1, 1995.

26. Tilders, F. J. H., DeRijk, R. H., Van Dam, A-M., Vincent, V. A. M., Schotanus, K., and Persoons, J. H. A., Activation of the hypothalamus-pituitary-adrenal axis by bacterial endotoxins: routes and intermediate signals, *Psychoneuroendocrinology,* 19, 209, 1994.

27. De Kloet, E. R., Vreugdenhil, E., Oitzl, M. S., and Joels, M., Glucocorticoid Feedback resistance, *Trends Endocrinol. Metab.,* 8, 26, 1997.
28. De Boer, S. F., Van Der Gugten, J., and Slangen, J. L., Plasma catecholamine and corticosterone responses to predictable and unpredictable noise stress in rats, *Physiol. Behav.,* 45, 789, 1989.
29. Jezova, D., Skultetyova, I., Tokarev, D. I., Bakos, P., and Vigas, M., Vasopressin and oxytocin in stress, *Ann. N. Y. Acad. Sci.,* 771, 192, 1995.
30. Mormede, P., Dantzer, R., Montpied, P., Bluthe, R-M., Laplante, E., and LeMoal, M., Influence of shock-induced fighting and social factors on pituitary-adrenal activity, prolactin and catecholamine synthesizing enzymes in rats, *Physiol. Behav.,* 32, 723, 1984.
31. Rivier, C., Luteinizing-hormone-releasing hormone, gonadotropins, and gonadal steroids in stress, *Ann. N. Y. Acad. Sci.,* 771, 187, 1995.
32. Schuurman, T., 1981. Hormonal correlate of agonistic behavior in adult male rats, *Prog. Brain Res.,* 53, 415, 1982.
33. Aguilera, G., Kiss, A., Luo, X., and Akbasak, B-S., The renin angiotensin system and the stress response, *Ann. N. Y. Acad. Sci.,* 771, 173, 1995.
34. Zukovska-Grojec, Z., Neuropeptide Y: a novel sympathetic stress hormone and more, *Ann. N. Y. Acad. Sci.,* 771, 219, 1995.
35. O'Donohue, T. L. and Dorsa, D. M., The opiomelanotropinergic neuronal and endocrine systems, *Peptides,* 3, 353, 1982.
36. Bodnar, R. J., Effects of opioid peptides on peripheral stimulation and 'stress'-induced analgesia in animals, *Crit. Rev. Neurobiol.,* 6, 39, 1990.
37. Morrow, L., Devaud, L. L., Purdy, R. H., and Paul, S. M., Neuroactive steroid modulators of the stress response, *Ann. N. Y. Acad. Sci.,* 771, 257, 1995.
38. Sternberg, E. M. and Licinio, J., Overview of neuroimmune stress interactions; implications for susceptibility to inflammatory disease, *Ann. N. Y. Acad. Sci.,* 771, 364, 1995.
39. Mizel, S. B., The interleukins, *FASEB J.,* 3, 2379, 1989.
40. Besedovsky, H. O. and Del Rey, A., Immune-neuro-endocrine interactions: facts and hypotheses, *Endocrin. Rev.,* 17, 64, 1996.
41. McCann, S. M., Lyson, K., Karanth, S., Gimeno, M., Belova, N., Kamat, A., and Rettori, V., Mechanism of action of cytokines to induce the pattern of pituitary hormone secretion in infection, *Ann. N. Y. Acad. Sci.,* 771, 386, 1995.
42. De Kloet, E. R., Brain corticosteroid receptor balance and homeostatic control, *Front. Neuroendocrinol.,* 12, 95, 1991.
43. Bohus, B. and De Wied, D., Pituitary-adrenal system hormones and adaptive behaviour, in *General, Comparative, and Clinical Endocrinology of the Adrenal Cortex, Vol. 3,* Chester-Jones, I. and Henderson, I. W., Eds., Academic, London, 265, 1980.
44. Bohus, B., Humoral modulation of learning and memory processes: physiological significance of brain and peripheral mechanisms, in *The Memory System of the Brain,* Delacour, J., Ed., World Science, Singapore, 337, 1994.
45. Kovacs, G. and De Wied, D., Peptidergic modulation of learning and memory processes, *Pharmacol. Rev.,* 46, 269, 1994.
46. McGaugh, J. L., Involvement of hormonal and neuromodulatory systems in the regulation of memory storage, *Ann. Rev. Neurosci.,* 12, 255, 1989.

47. Roozendaal, B., Cahill, L., and McGaugh, J. L., Interaction of emotionally activated neuromodulatory systems in regulating memory storage, in *Brain Processes and Memory,* Ishikawa, K., McGaugh, J. L., and Sakata, H., Eds., Elsevier, Amsterdam, 39, 1996.

48. Rose, S. P., Cell adhesion molecules, glucocorticoids and long-term memory formation, *Trends Neurosci.,* 18, 502, 1995.

49. Bohus, B., Neurohormones, synaptic plasticity, and memory, *Int. Rev. Cytol.,* 1998.

50. Ader, R. and Cohen, N., The influence of conditioning on immune responses, in *Psychoneuroimmunology, 2nd ed.,* Ader, R., Felten, D. L., and Cohen, N. Eds., Academic Press, San Diego, 1991.

51. Pavlov, I. P., *Conditioned Reflexes,* Dover, New York, 1927.

52. LeDoux, J. E., Emotion: clues from the brain, *Ann. Rev. Psychol.,* 46, 209, 1994.

53. Ader, R., Conditioned adrenocortical steroid elevation in the rat, *J. Comp. Physiol. Psychol.,* 90, 1156, 1976.

54. Coover, G. D., Sutton, B. R., and Heybach, J. P., Conditioned decrease in plasma corticosterone level by pairing stimuli with feeding, *J. Comp. Physiol. Psychol.,* 91, 716, 1977.

55. Graham, J. M. and Desjardins, C., Classical conditioning: Induction of luteinizing hormone and testosterone secretion in anticipation of sexual activity, *Science,* 210, 1039, 1980.

56. Roozendaal, B., Koolhaas, J. M., and Bohus, B., Central amygdaloid involvement in neuroendocrine correlates of conditioned stress response, *J. Neuroendocrinol.,* 4, 483, 1992.

57. Korte, S. M., Buwalda, B., Bouws, G. A. H., Koolhaas, J. M., and Bohus, B., Conditioned neuroendocrine and cardiovascular stress responsiveness accompaning behavioural passivity and activity in aged and in young rats, *Physiol. Behav.,* 51, 815, 1991.

58. Van de Kar, L. D., Piechowski, R. A., Rittenhouse, P. A., and Gray, T. S., Amygdaloid lesions: Differential effect on conditioned stress and immobilization-induced increases in corticosterone and renin secretion, *Neuroendocrinology,* 54, 89, 1991.

59. Gray, T. S., Piechowski, R. A., Yracheta, J. M., Rittenhouse, P. A., Bethea, C. L., and Van de Kar, L. D., Ibotenic acid lesions in the bed nucleus of the stria terminalis attenuate conditioned stress-induced increases in prolactin, ACTH and corticosterone, *Neuroendocrinology,* 57, 517, 1993.

60. Onaka, T., Palmer, J. R., and Yagi, K., Norepinephrine depletion impairs neuroendocrine responses to fear but not novel environmental stimuli in rat, *Brain Res.,* 713, 261, 1996.

61. Gray, T. S., Amygdaloid CRF pathways: role in autonomic, neuroendocrine, and behavioral responses to stress, *Ann. N. Y. Acad. Sci.,* 697, 53, 1993.

62. Loewy, A. D., Forebrain nuclei involved in autonomic control, *Progr. Brain Res.,* 87, 253, 1991.

63. Bohus, B., Koolhaas, J. M., Luiten, P. G. M., Korte, S. M., Roozendaal, B., and Wiersma, A., The neurobiology of the central nucleus of the amygdala in relation to neuroendocrine and autonomic outflow, *Progr. Brain Res.,* 107, 447, 1996.

64. Sapolsky, R. M., Social subordinance as a marker of hypercotisolism: some unexpected subtleties, *Ann. N. Y. Acad. Sci.,* 771, 626, 1995.

65. Koolhaas, J. M., Hermann, P. M., Kemperman, C., Bohus, B., Van den Hoofdakker, R. H., and Beersma, D. G. M., Single social defeat in male rats induces a gradual but long lasting behavioural change: a model of depression? *Neurosci. Res. Comm.,* 7, 35, 1990.

66. Van Dijken, H. H., De Goeij, D. C., Sutanto, W., Mos, J., De Kloet, E. R., and Tilders, F. J., Short inescapable stress produces long lasting changes in the brain-pituitary-adrenal axis in adult male rat, *Neuroendocrinology,* 58, 59, 1993.

Chapter 11

Issues in Quantifying Pulsatile Neurohormone Release

Johannes D. Veldhuis

Contents

I. Introduction/Overview

Neuroendocrine data are particularly challenging to analyze quantitatively in view of their unique characteristics, such as the tendency of observations to be relatively infrequent (i.e., short or coarse data series), noisy, confounded by variable half-lives and nonuniform experimental uncertainty, and composed of unknown true signal (e.g., theoretical waveform of secretion over time), with or without concomitant basal hormone release.[1,2] Since neuroendocrinology is concerned with quantal signaling and interglandular communication, quantification of short-term variations and longer-term trends in neurohormone measurements are often critical to investigating and understanding mechanisms of *in vivo* or *in vitro* neurohormone regulation, pathophysiological disturbance, gender differences, etc. Thus, a primary task of neuroendocrine investigation is accurate, reproducible, valid and vivid quantitation of contrasts between and among neurohormone time series. This goal is made more difficult by potential background drift in the data mean or variance (technically termed, nonstationarity), differences in some circumstances between decay rates of exogenously injected and endogenously secreted hormone, unequal distribution spaces for various neurohormones (e.g., testosterone approximately 32 liters in a 70-kilogram adult, whereas luteinizing hormone (LH) released within the male reproductive axis is confined largely to the 3.5 liters of plasma volume), as well as the necessity to obtain precise, sensitive, specific, reliable, and valid neurohormone measurements. Indeed, the last mentioned issue, while beyond the scope of this review, represents a critical step in quantitative neuroendocrinology.[3] Last, neurohormone rhythms commonly occur on multiple time scales concurrently, e.g., ultradian with events occurring multiple times over any given 24-hr period, circadian (approximating a 24-hr rhythm), and sometimes longer-term trends that extend over weeks or a month (e.g., menstrual cycle). Even more extended trends exist that are seasonal or circannual. Given these issues in neuroendocrine data acquisition, processing, and analysis, pulsatile or episodic neurohormone release cannot be evaluated adequately without thoughtful, pertinent, and intensive attention to a variety of primary issues. Here, I will review several distinct issues in quantitative neuroendocrinology, including the specific quantification of pulsatile neurohormone data, evaluation of subordinate (non-pulsatile) patterns within the measurements, the notion of feedback and feed-forward control (network function of a neuroendocrine axis), the potential concordance or synchrony in pulse trains, and approaches to establishing reproducibility of biological data. Of necessity, some issues remain incompletely resolved, and these will be noted.

II. Primary Issues

A. Pulsatile Features of Neuroendocrine Data

In signaling proximal or distal target tissues, neuroendocrine ensembles typically release quantal signals best described as bursts or pulses of hormone secreted into a relevant compartment, often the bloodstream but also the cerebrospinal fluid, etc. Thus, serial measurements of neurohormone concentrations over time typically yield sharply varying values characterized by abrupt increases and decreases of nonuniform magnitude, which can be defined as pulsatile events. In the clinical neurohormone arena, pulses of LH and GH release were first recognized approximately 25 years ago. Episodic peaks confound data gathering and analysis by requiring multiple observations in order to obtain accurate mean and integrated estimates. Indeed, a hallmark of appropriate data quantitation is to plot and visualize the serial neurohormone measurements over a given time window, in order to gain some appreciation of the volatility (extent of pulsatile release) as well as longer-term trends (e.g., 24-hr rhythms) in the data. This issue is illustrated in Figure 1, which shows serum LH concentrations measured in blood samples collected at frequent intervals in a healthy young adult. In the human, LH release exhibits 12 to 20 macroscopic "pulsatile" events per 24 hr, in which serum LH concentrations increase by 1 to 3 fold over several minutes, and then decay toward interpulse baseline values over an additional 90 to 200 min.[4] The quantification of pulsatile LH release has disclosed that in the female LH secretory burst amplitude, duration, mass, and frequency vary significantly, whereas the half-life and daily secretion of LH are less substantially altered throughout a 30-day menstrual month.[5] In addition, 24-hr trends within serum LH concentration profiles exist, such that nighttime mean LH concentrations may rise as the pulse frequency declines but the amplitude of individual bursts increases.[6] Although inspection of Figure 1 demonstrates distinct punctuated pulsatile episodes of LH release, many neuroendocrine data do not exhibit this organized release pattern, thus requiring objective, reproducible, and valid methods for peak analysis.

B. Discrete Peak Detection

Discrete peak detection methods have emerged since the early 1970s, as algorithms designed to identify and quantitate the amplitude and frequency of abrupt neurohormone concentration changes over time; i.e., to detect discrete pulses or peaks in the data. Many of the available methods are relatively model-free, as reviewed in detail elsewhere.[4] For example, one strategy entails computer-assisted serial testing of putative nadir and peak clusters of neurohormone measurements to mark statistically significant increases (peak upstroke) in the data, followed then by a corresponding search for all significant decreases (peak downstroke) according to some relevant threshold or statistical criterion.[7] The combination of a significant up-stroke followed by down-stroke in the data is taken to represent a pulse or peak. This is illustrated by way of so-called Cluster analysis in Figure 2.

In general, discrete peak detection models require knowledge of the inherent experimental uncertainty within the assay, and ideally also within the preassay sample withdrawal and processing steps. Significant changes in neurohormone concentrations are then judged in relation to this background variability. In general, these methods should require a 2 to 3 fold increase in signal over background variability in order to restrict type I or false-positive statistical errors. On the other hand, excessively strict criteria for peak identification will predispose to so-called type II or false-negative statistical errors, in which real pulse events within the data are overlooked.[8-10]

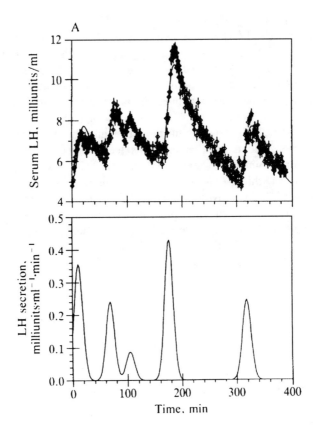

FIGURE 1

Repetitive sampling of blood LH concentrations every 1 min in a normal man. Macroscopic episodes of pulsatile LH (RIA) release are evident from visual inspection of the serum LH concentrations (upper panel) and by computer-assisted deconvolution analysis of underlying secretory rates (lower panel). Bursts of secretion are scattered seeming at random intervals. (Adapted from Veldhuis, J. D. et al., *Proc. Natl. Acad. Sci. USA*, 84, 7686, 1987. With permission.)

Methods of discrete peak detection that allow for some drift in baseline are also important, in view of the 24-hr rhythmicity or short-term trends that many neurohormone profiles exhibit.[11] Thus, for example, the Cluster algorithm makes repeated ("moving window") local tests for significant increases and decreases in the data, and thus is relatively less influenced by slow longer-term trends. Alternatively, Pulsar defines a nominally smoothed baseline in the data over a particular (several-hour) window length, and evaluates significant increases in relation to that calculated baseline.[12] Other discrete peak detection methods such as Detect, Cycle Detector, Ultra, etc., have been reviewed and compared recently.[4] All these methods evaluate pulsatile changes in blood neurohormone *concentrations*, whereas only some of the methods attempt to calculate underlying neurohormone *secretory* rates, e.g., by deconvolution analysis of the plasma hormone concentrations (see below). Deconvolution analysis practically allows one to appraise the apparent underlying secretory

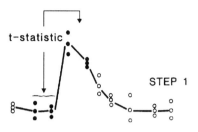

Cluster Analysis (1986)

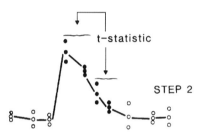

Cluster Analysis (1986)

FIGURE 2

Schematized presentation of concept of discrete neurohormone peak detection via Cluster analysis.[7] The Cluster algorithm is designed to search serial hormone measurements sequentially for the occurrence of any significant abrupt increase in mean concentration values, as judged via a pooled t statistic applied to a moving test nadir and peak. All significant increases in the data are marked in the first scan of the series (step 1). Thereafter, significant decreases are identified in the data (step 2). Last, peaks are defined as regions preceded by a significant increase and concluded via a significant decrease. (From Urban, R. J. et al., *Endo. Rev.,* 9, 3, 1988. ©The Endocrine Society. With permission.)

activity of a gland, and technically decreases the autocorrelation in the concentration data, by removing the confounding impact of (variable) hormone half-life on the pulse profile.

C. Deconvolution Analysis of Secretion

Deconvolution analysis encompasses a class of mathematical techniques designed to extract from the neurohormone concentration data estimates of underlying hormone secretion rate and/or half-life.[1,2,4,13-29] In general, these methods have been developed more recently than discrete peak detection algorithms, and are useful to approximate underlying neurohormone secretion profiles under one or more relevant assumptions. Most deconvolution analysis algorithms require an assumed or *a priori* known neurohormone half-life. An early version of deconvolution analysis employed in the Detect program, for example, uses a one or two component hormone half-life, which is obtained from the literature or in independent experiments, to calculate the amount of hormone secretion that would occur between successive blood samples in order to generate the serum neurohormone concentration observed at each time.[30] This procedure has been referred to as calculating the instantaneous (sample-by-sample) secretion rate (ISR). Whereas this methodology is quite model-free, at high sampling intensities (e.g., blood withdrawal every 1 to 5 min), the mathematical construction of the ISR calculation is ill-conditioned, resulting in oscillatory ("jerky") secretion rates that may become negative and not faithfully represent the prolonged secretory episode.[4] Appropriate strategies have been developed more recently to limit the ill-conditioning of the deconvolution problem, including smoothing techniques as well as non-negativity constraints, as reviewed recently by de Nicolao.[14]

Model-specific deconvolution algorithms have also been proposed, in which the secretory event is envisioned as a smooth distribution of secretion rates centered about some instant in time.[28,29,31] This concept, as employed in so-called multiparameter deconvolution analysis (DECONV), is illustrated in Figure 3. In this particular deconvolution technique, putative neurohormone release episodes are approximated algebraically by Gaussian distributions of instantaneous release rates, which can occur at randomly spaced intervals throughout the observation session. In this "burst model" concept, each neurohormone molecule released into the bloodstream is acted upon by some hormone-, subject- and condition-specific half-life, which can be estimated simultaneously with secretion by solving the so-called convolution integral equations. These equations relate each measured serum neurohormone concentration to all prior neurohormone secretory pulses as well as the relevant neurohormone half-life.[28] For any one data set, a collection of equations (one equation for each sample hormone value) is then evaluated iteratively for unknown secretory burst locations, amplitudes, and durations as well as neurohormone half-life. This seemingly formidable task in computation is achievable with modern high-speed microcomputers, and allows model-based estimates of hormone secretion and half-life simultaneously. The application of this method is illustrated in the lower panel of Figure 1.

An assumption of the multiparameter methodology is that a neurohormone release episode can be approximated reasonably by a Gaussian distribution of release rates.[31] The validity of this assumption in DECONV is suggested by recent sampling of the human inferior petrosal sinus, which directly receives pituitary venous drainage, where serum LH and GH concentrations were measured in blood obtained in frequent, e.g., every 5 min, intervals. As shown in Figure 4, the release profiles of LH over time in different patients can be approximated by Gaussians of variable half-width and amplitude. In addition, the simple Gaussian model predicts the mean neurohormone half-life for multiple hormones relatively well, estimates the mean endogenous secretion rate comparably to general literature values, fits most data profiles with a high degree of goodness of fit and randomly distributed residuals, and has found application in a host of neuroendocrine pathophysiological states. In addition, a novel waveform-independent deconvolution method (PULSE) yields computed hormone neurosecretory episodes that are approximately symmetric, and typically superimposed upon little or no basal hormone release.[29] For example, we have recently also utilized PULSE to analyze secretion, whereby the serum hormone

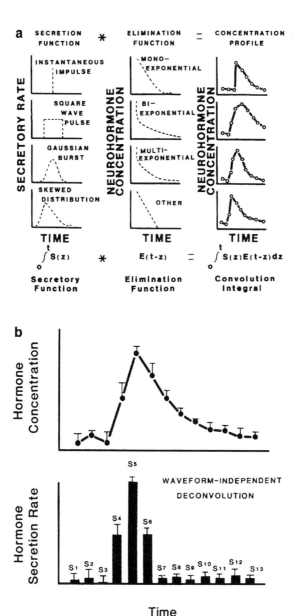

FIGURE 3

Concept of waveform-specific and waveform-independent deconvolution analysis. *Panel A* summarizes the concept of deconvolution analysis, by which a serum hormone concentration peak (right-most) is decomposed mathematically into its constituent secretory burst (e.g., a Gaussian or other secretory rate distribution) as well as the corresponding hormone-specific decay process (e.g., monoexponential removal rate). *Panel B* shows a model-independent deconvolution method, by which one can calculate the array of apparent sample secretion rates (middle panel) that result in a serum neurohormone concentration pulse (lower panel), given a known or well estimated one or two-component hormone half-life.[29] The two approaches to deconvolution analysis are complementary, allowing for different starting assumptions and implicit prior knowledge. (Adapted from Veldhuis, J. D. and Johnson, M. L., *Meth. Enzymol.*, 210, 539, 1992. ©The Endocrine Society. With permission.)

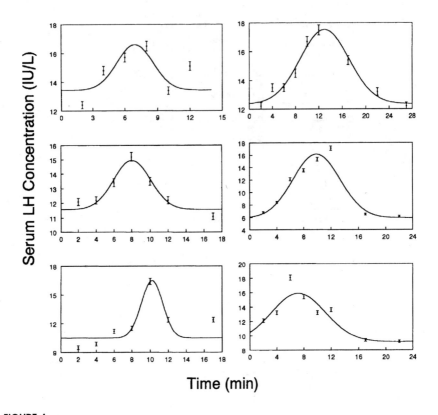

FIGURE 4

Time course of LH release defined by sampling the inferior petrosal sinus of men and women for later assay of luteinizing hormone (LH). The data shown derived from 6 separate individuals, and the plotted curves denote a fitted Gaussian distributions of the hormone concentrations over time. Data are from Professor Lombardi and Dr. A. Colao (Naples, Italy) and JD Veldhuis (unpublished).

concentrations over time are assumed to result from a corresponding particular array of sample secretion rates each influenced by a known (relevant) two-component half-life of hormone disposal. With sensible positivity constraints (e.g., solving for the logarithm of the amplitude rather than the amplitude itself, and using an approximating deconvolution technique initially described in spectroscopy), hormone profiles can be deconvolved into model-free estimates of sample secretion rates with error estimates.[29] These secretory profiles can then be analyzed further, as desired, e.g., via Cluster, Pulsar, etc., for basal vs. pulsatile release.

We have termed a still more recent version of model-free deconvolution analysis PULSE2. In this strategy of deconvolution analysis (reviewed in detail in Reference 1), placement of possible individual secretory impulses throughout the hormone profiles is iteratively evaluated, assuming a particular known mono- or bi-exponential hormone half-life. Each secretory impulse that results in a significantly reduced fitted variance by F ratio testing for the data series, i.e., a better fit than its predecessor, is retained, at some particular *a priori* probability value (e.g., $p < 0.1$,

0.05, or 0.01, etc.). This methodology is sensitive to choice of the smoothing window employed, and also appropriately to the probability limits imposed for retaining or rejecting a putative peak. However, PULSE2 is relatively insensitive to half-life specification, resulting in good estimates of pulse number and amplitude over a range of nominal half-lives.[1] Additional validation and other novel approaches to waveform-specific and waveform-independent deconvolution analyses will be important to aid in accurate and robust quantitation of neuroendocrine pulse activity. Such methods ultimately might include nonparametric, neural-network or fuzzy-logic based approaches, etc.

D. Relationship Between Ultradian and Circadian Rhythms

In inspecting a 24- or 48-hr serum neurohormone concentration profile, one typically observes both ultradian (pulsatile) and longer-term 24-hr rhythms (e.g., sleep activity and/or circadian cues). For example, in the ACTH-cortisol axis, mixed ultradian and circadian rhythmicities are immediately evident.[32,33] An important mechanistic issue that can be addressed by deconvolution analysis is how ultradian episodic secretion relates to circadian variations in neurohormone concentrations in the circulation.[32,33] With appropriate waveform-specific as well as waveform-independent deconvolution analyses, we have observed that the daily rhythm in plasma ACTH concentrations is explained almost exclusively via a 3- to 4-fold variation in ACTH secretory burst mass over the course of 24 hr.[32] In contrast, the circadian variation in serum cortisol concentrations develops in large part from a variation in cortisol secretory burst mass, and in smaller measure from a day/night difference in cortisol secretory burst frequency or interpulse interval.[33] Growth hormone release also shows marked 24-hr variations in secretory burst mass (up to 30-fold differences across the day and night), and to a lesser degree (1.5 to 3 fold) in secretory burst frequency.[34] Deconvolution analysis thus provides one tool with which to address the question, "How are circadian rhythms assembled from underlying short-term pulsatile activities?".

E. Deconvolution Analysis of Physiologically Paired Hormone Series

Some hormones are co-secreted; e.g., insulin and C peptide in equimolar ratio, and LH and FSH in relative quantitative synchrony.[35,36] Appropriate pulse analysis ideally would be founded in these circumstances in part on expected co-secretion of the hormones, e.g., ACTH and beta endorphin. To this end, in the arena of insulin pathophysiology, deconvolution methods have utilized C peptide measurements to calculate insulin secretion, based on the well determined half-life of C peptide in peripheral blood, and the physiological precept that it is co-secreted with insulin.[37] We have recently modified a multiparameter deconvolution analysis method (DECONV, above) to allow simultaneous estimates of hormone secretion from two relevantly paired series, e.g., LH and FSH; C peptide and insulin, etc. We have not yet modified PULSE or PULSE2 in this manner. By recognizing the physiological coupling between two hormone secretion profiles, and by measuring concentrations in both series, additional statistical power is achieved in estimating their common underlying secretory rates, after appropriate adjustment for their dissimilar half-lives.[2]

F. Subordinate Release Patterns in Neurohormone Data

In addition to overtly pulsatile neurohormone release, minute-to-minute variations in neurohormone concentrations occur in the interpulse regions as well as during the decay portions of pulses, and even in their ascending limbs. The ability to evaluate the relative orderliness or regularity of such subordinate changes in the data has been very limited until recently. For example, classical methods of nonlinear dynamical systems that assess deterministic chaos in data typically require 100,000 or more observations in order to calculate the rate of entropy, correlation dimension, Lyapanov exponents, etc. In contrast, a recently developed and novel statistical, approximate entropy (ApEn), is useful for data series as short as 50 to 100 observations.[38] ApEn provides a scale- and translation-invariant statistic, as well as a model-independent measure, of the regularity or orderliness of sub-pattern reproducibility within serial measurements, e.g., neurohormone concentrations or secretion over time. Normalized ApEn adjusts for differences in absolute hormone concentrations by testing for pattern reproducibility by way of a threshold defined by some fraction of the standard deviation of the overall neurohormone profile. In addition, by first-differencing of the data (calculating the algebraic differences between successive measurements), long-term trends in the profile can be obviated, allowing the quantitation of subordinate patterns independently of initial nonstationarity (i.e., varying mean or variance of the series across the observation interval).

ApEn is a family of statistics, any given member of which is defined by a particular M, R, and N. M represents the "window length", or the length of the vector defining a potential pattern of recurrence in the data. R represents the tolerance or threshold, i.e., the range within which the vector length must be repeated in order to demonstrate orderliness. N designates the total number of data points. As the value of M or N increases, the absolute value of ApEn rises, and thus ApEn is best defined at any particular M, R, and N. Of importance statistically, ApEn is asymptotically normally distributed, and is a positive real number for both stochastic and/or deterministic data series, which can distinguish between *degrees* of randomness. Thus, unlike classical entropy measures which can tend to infinity in the presence of low constant amounts of noise, or purely stochastic processes, ApEn is finite, real, positive, and non-zero with a standard deviation in neurohormone data analysis of 6 to 15% of its value.

As reported earlier, ApEn estimates of the regularity of GH or ACTH release provide sharp discrimination between tumoral GH or ACTH secretion and that of normal individuals, by showing consistently greater randomness for tumoral hormone release.[39,40] Moreover, ApEn detects male-female differences in GH release patterns[41] and subtle disorganization within GH[34] as well as LH and testosterone time series of aging individuals, even when the mean serum LH (or testosterone) concentration is no different from that of young adults: Figure 5.[42] Thus, ApEn evaluates subordinate patterns within neurohormone profiles, and is complementary to conventional pulse analysis. Indeed, under conditions of tumoral hormone hypersecretion, pulse analysis can be very difficult given sparingly reproducible waveforms ("irregularly irregular" pulses) in the data.

F. Synchrony Between Neurohormone Secretory Pulse Trains or Concentration Profiles

Neurohormone systems often operate with temporal delays and strong mechanistic linkages, e.g., the synchronous release of LH and FSH, ACTH and beta endorphin, and C peptide and insulin (see above) from the same secretory granules. In addition, the nature of neuroendocrine axes includes feed-forward and feedback relationships, typically governed by dose-response functions and appropriate physiological lags. This is well illustrated in the GnRH-LH-testosterone (male reproductive) axis, and in the CRH-ACTH-cortisol adrenal stress-responsive axis. In evaluating the overall behavior of these neuroendocrine systems, it is often important to quantify the degree of synchrony between relative hormone pairs. Consequently, quantitative methods to evaluate the synchrony, correlation, or co-pulsatility of neurohormone time series are important.

Cross-correlation analysis represents the classical approach to quantifying directionally similar (positive cross-correlation) or directionally opposite (negative cross-correlation) between serial measures of (usually equally spaced) neurohormone concentrations, e.g., ACTH and cortisol, LH and testosterone, etc.[43] One strength of the cross-correlation analysis is its independence from any assumption about pulse analysis, and its statistical power when data series are relatively extended. The statistical power derives from the approximation that the standard deviation of the cross-correlation coefficient is inversely proportionate to the square root of N-k, where N denotes the number of data points, and k the so-called lag. Cross-correlation can be carried out at various lags for any given hormone pair; i.e., by considering different times separating the LH and testosterone measures, such as allowing testosterone to lag LH by 10, 20, or 30 min. For a neurohormone sampled every 10 min for 24 hr, where N = 145, and a lag of 1 (e.g., testosterone measures are correlated with LH values 10 min earlier), the nominal standard deviation of the cross-correlation coefficient is approximately 0.083, which indicates that cross-correlation (r) values in excess of 0.16 (twice r) are unlikely to be due to chance alone. More recently, we have also used Monte Carlo methods to evaluate the empirical standard error of r, which is accomplished by shuffling each hormone time series randomly and recalculating r several hundred times for the scrambled series. These r values reflect essentially random cross-correlation coefficients, whose mean value should approximate zero with some finite standard deviation.

Unfortunately, cross-correlation analysis can be confounded by variable lags within the data[42] and by significant autocorrelation within each hormone profile.[43]

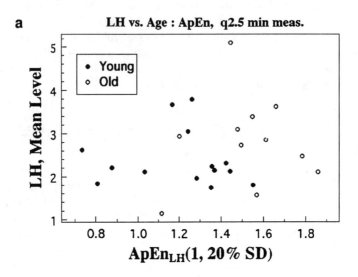

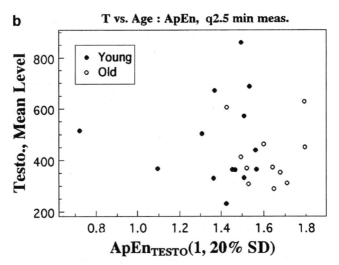

FIGURE 5

Approximate entropy (ApEn) analysis of serum LH and testosterone profiles collected by sampling blood every 2.5 min overnight in a group of young and older men. Panel A gives the scatterplot of mean serum LH concentrations against ApEn values, illustrating a sharp contrast in ApEn but not in mean serum LH concentrations as a function of age. Panel B illustrates the corresponding serum concentrations and ApEn values for testosterone release overnight. Note that higher absolute ApEn values occur for both LH and testosterone in older men. Such values denote greater disorderliness of the release process. (Adapted from Pincus, S. M. et al., *Proc. Natl. Acad. Sci. USA*. With permission.)

The latter is the tendency of any given serum hormone concentration to predict its predecessor and its successor, given that the hormone has a finite and definite half-life in the blood. Cross-correlation analysis with autoregressive modeling is useful to remove this impact of autocorrelation on the cross-correlation coefficient. *Alternatively, we suggest that the target-gland neurohormone concentration profile be deconvolved to generate its underlying secretion rates instead.* One then cross-correlates the effector hormone concentration values and the calculated target gland secretion rates. This eliminates metabolic-clearance related autocorrelation at least in this series. On the other hand, it is less relevant to deconvolve the effector hormone concentration profile (e.g., ACTH or LH, driving cortisol or testosterone, respectively), since the target gland is believed to be driven in proportion to the plasma concentration (rather than in proportion to the trophic hormone secretion rate *per se*). These issues are discussed in further detail elsewhere in a chapter devoted to neurohormone synchrony.[43]

A second strategy that is conceptually distinct and mathematically independent is first to identify discrete pulses within the two neurohormone profiles, and then ask the question whether there is concordance of individual pulses (concurrence or co-pulsatility) to an extent exceeding that expected on the basis of chance alone. In this circumstance, one needs to estimate the random concordance given any particular pulse frequencies in the pair of independent neurohormone pulse trains.[10,36] This value can be estimated from the hypergeometric probability distribution, and represents the product of the fractional pulse numbers in the two series, e.g., given blood sampled every 10 min for 24 hr (145 samples), and assuming series A and B containing respectively 20 and 15 hormone pulses, the expected concordance rate on the basis of chance alone is $20 \times 15 \div 143$ (2 is subtracted from 145, since peaks cannot be identified definitively in the first and last samples of a time series). In addition, one can calculate the standard deviation of the mean number of pulses expected on the basis of chance alone, as well as compute the probability density. A PC-based software program is available to make such calculations, and to estimate the probability that at least the observed number of coincidences has occurred on the basis of chance alone, i.e., a P value of importance to the investigator.[36] Of note, should there be significant diurnal variation in hormone pulse frequency, then neurohormone series can be subjected to Monte Carlo pseudo-randomization of peak locations, and the paired randomized series evaluated for chance concordance rates. Where many observed neurohormone pulse trains are available, one can also randomly assign paired comparisons in order to estimate concordance rates for irrelevant pairs.[43]

In addition to cross-correlation analysis and discrete coincidence calculations, the cross-approximate entropy (cross-ApEn) statistic is distinct in that it allows one to evaluate conditional *orderliness* of two hormone series.[42] Cross-ApEn calculates the conditional regularity of two discretized profiles, by quantifying the degree to which a pattern represented in the first neurohormone series is reflected sooner or later in the second. Importantly, cross-ApEn is calculated independently of lag, and therefore has statistical power to define nonuniformly lagged repetition of patterns across two measured series. This stands in contrast to cross-correlation (or cross-spectral) analysis, wherein a relatively fixed lag is required for good statistical power to identify correlations. Thus, cross-ApEn was able to delineate strong statistical differences in pattern reproducibility between overnight LH and testosterone profiles in older men compared to younger individuals, whereas cross-correlation analysis did not reveal this extent of contrast: Figure 6.[42]

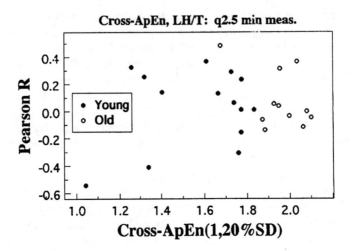

FIGURE 6

Cross-approximate entropy (cross-ApEn) as a statistic to quantify conditional orderliness or relative regularity of paired hormone series, here LH and testosterone, in young compared to older men. Volunteers were sampled as described in the legend of Figure 5, and cross-ApEn values were computed to quantify relative orderliness or synchrony of the LH and testosterone release. All older men except one exhibited higher cross-ApEn values than younger controls, denoting diminished synchrony between LH and testosterone release with aging. (Adapted from Pincus, S. M. et al., *Proc. Natl. Acad. Sci. USA*. With permission.)

Thus, the three methodologies described briefly above allow one to pose and address different experimental and clinical questions. They are complementary and not mutually exclusive.

III. Reproducibility of Neuroendocrine Quantitation

Biological test-retest reliability of neurohormone profiles is important to establish. For example, in the case of pulsatile LH and GH release, we have observed that the 24-hr mean serum concentration value is well predicted from prior 24-hr means in the same individual, and less so from the cohort as a whole, i.e., there is approximately 2- to 3-fold greater between-subject than within-subject variability of the 24-hr serum LH and GH mean concentration in men and women.[44-46] In addition, we have evaluated the reproducibility of deconvolution-based pulse analysis of LH profiles in young men, by sampling a cohort of 10 young men each on 3 separate occasions for 24 hr (each sampling session separated by at least 2 weeks). Here, we observed that our multiparameter-deconvolution estimates of LH half-life and the calculated daily production rate varied by less than 15%.[44] In contrast, large coefficients of variation define within group (between-individual) variability for neurohormone measurements in most circumstances. For example, as shown for the GH

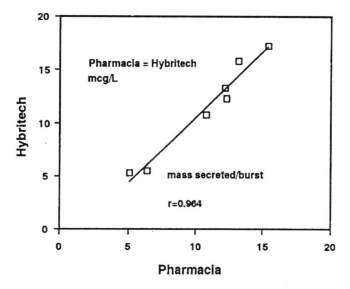

FIGURE 7

Comparison between two different immunoradiometric assays (IRMA) for GH, as applied to 24-hr serum GH concentration profiles deconvolved by multiparameter deconvolution analysis to estimate GH secretory burst mass in a group of boys (Schoenberg and Veldhuis, unpublished). A high degree of reliability or inter-test reproducibility is a necessary but not sufficient feature of valid neuroendocrine pulse analysis.

axis in prepubertal boys, the daily GH secretion rate is better determined upon serial re-sampling of the same individual over 24 hr, compared to between subjects.[45] In addition, the half-life of GH is relatively well estimated with a coefficient of variation of less than 15% within individuals, but shows a larger biological variation between subjects. We recommend in any new neurobiological investigation that the reproducibility of the neurohormone profile be evaluated both within and between subjects.

An additional way to evaluate reproducibility is to redo the assay of the hormone time series, and reanalyze the data. The assay may be repeated in a different configuration, e.g., with different antibodies, or simply repeated in its original form to assess measurement precision. As shown in Figure 7, using two independent immunoradiometric assays for GH applied to 20-min blood sampling over 24 hr, followed by deconvolution analysis of the two resultant GH profiles for each subject, we find a high correlation between the two separately calculated values of the mass of GH secreted per burst. Such reproducibility is essential to establish when interpreting clinical and experimental data.

Reproducibility of the analytical tool between users and on successive users should also be demonstrated, which in the case of multiparameter deconvolution analysis exceeds 97%.[47,48] Where possible, the entire algorithm should be fully automated to reduce the necessity for operator interactions with the program decisions, as in the case of PULSE, PULSE2, Cluster, Detect, etc.

III. Unresolved Issues

Many unresolved issues remain that are under active investigation in the field of neuroendocrine pulse analysis. For example, a major challenge entails distinguishing a model of basal neurohormone release with superimposed pulsatile events from an alternative model of nearly confluent (i.e., partially overlapping) secretory bursts without interpulse basal release. This problem is discussed in detail elsewhere, and represents an unresolved issue.[49] We suggest that physiological experiments may be helpful in making this distinction, e.g., by slowing the pulse frequency in the axis under study, in order to evaluate whether interpulse basal neurohormone release actually occurs. Such interventions may be essential to discriminate definitively between a model of admixed basal and pulsatile release vs. rapidly successive pulses within incomplete decay to baseline. For example, we have observed that measured total serum testosterone concentrations never fall to zero in blood sampled at 10-min intervals over 24 hr in a young man deprived of testosterone via ketoconazole treatment to block testicular steroidogenesis, and then infused with 0.5 mg testosterone base i.v. every 90 min to recapitulate pulsatile testosterone secretion.[50] Rather, serum testosterone concentrations rise rapidly to very high values immediately following i.v. bolus testosterone injection, and then fall toward apparent pseudosteady state ("interpulse basal release") between injections. In the absence of prior knowledge, this pattern could be interpreted as basal testosterone infusion combined with pulsed testosterone injection, whereas the experimental paradigm consisted solely of pulsed injection thus refuting this interpretation. This model illustrates the challenge of separating admixed basal and pulsatile hormone release from partially overlapping secretory bursts, especially in the presence of (a) binding protein(s) in blood.[51] The presence of sex-hormone binding globulin in blood likely causes the interpeak valley levels observed here.

Whether endogenously secreted and exogenously injected hormone half-lives are identical is not widely discussed in the literature, but represents an important assumption for many deconvolution methods. This may be especially important when multiple endogenous hormone isoforms exist. In addition, we have recently observed that the temporal mode of delivery of the identical hormone into the bloodstream can greatly alter its apparent half-life.[52] For example, in the presence of the high-affinity GH binding protein in plasma, GH kinetics are non-identical following bolus vs. more prolonged infusion. Indeed, under octreotide treatment to suppress endogenous GH release, we can estimate the half-life of recombinant human GH injected i.v. by bolus as approximately 8.5 to 10 min in healthy young men and women. When recombinant GH is infused at steady-state, the apparent half-life in plasma approaches 18 to 20 min. Moreover, if the infusion of GH at steady-state is abruptly terminated, and the half-life of decay of GH toward baseline computed directly, the directly calculated half-life of is 20 to 25 min. By comparison, deconvolution analysis of endogenous GH release suggests a half-life of approximately 18 to 20 min.[53] Accordingly, the apparent half-life of hormone removal can vary as a function of its mode of entry into the bloodstream, as well as in some cases in relation to the mean hormone concentration achieved, the presence of high affinity binding proteins in plasma, the ability of the liver and kidney to remove bound vs. free hormone, etc.[54] This area requires further analyses and experimental investigation.

Of further note, as shown in Figure 8, fitting a single half-life for variable concentration pulses, e.g., as induced by different doses of i.v. GnRH in patients with congenitally GnRH deficiency, cannot always be appreciated from simple visual inspection of the hormone profiles.[55] Similarly, consider a 12-hr LH profile consisting of an 8-hr baseline followed by two GnRH injections.[56] The *apparent* slopes of the decay of serum LH concentrations after the variously large and small peaks may seem to differ. However, a single half-life fits all the pulses, because the decay portions contain unequal amounts of continuing secretion depending on their different amplitudes. The combined contributions of variable secretion rates and a common half-life in all peaks is addressed by deconvolution analysis.

A neuroendocrine axis is subject to agonist and antagonist dose-response inputs at several regulatory loci; e.g., in the case of the reproductive axis, the agonistic GnRH-driven LH secretory response (pituitary); the stimulatory LH-directed testosterone secretory response (testis); and the inhibitory testosterone-LH negative feedback dose-response (hypothalamus). In general, virtually no information is available regarding the *in vivo* characteristics of these "hidden" dose-response curves under unperturbed and uninterrupted conditions; i.e., without isolating or clamping a component of the system experimentally. A suitable physiologically structured network of functional nodes with relevant dose-response curves to describe the overall dynamics of an axis should eventually allow estimation of endogenous dose-response functions in health and disease. This problem is just beginning to be addressed, and remains an important long-term challenge. We predict that altered hormone dose-response functions and elimination rates will exist in different physiological conditions as well as in disease states.

Non-parametric pulse detection methods will also become useful, as they are developed. These approaches should be complementary to existing methods, since assumptions regarding the waveform of the secretory impulse, and regarding the distribution of experimental uncertainty in the data, cannot always be evaluated experimentally. Much more progress is needed in this area.

The monitoring of long-term neuroendocrine axis stability over days has been accomplished only to a very limited extent, in view of current assay sensitivity and blood volume restrictions. However, if blood is sampled at frequent (e.g., 20 min) intervals over several consecutive days, the overall stability of the axis would become evident; e.g., in its ability to reproduce relatively stable circadian rhythms and ultradian pulsatile activity. As neuroendocrine tools of sampling and assay improve, longer-term studies of daily and weekly hormone output will become important in order to understand mechanisms of feedback control over longer intervals. For example, sampling throughout the entire menstrual (month-long) cycle will be important eventually to define an array of relevantly interacting hormones.

Very little is known about tumoral mechanisms that disrupt hormone secretion. However, as noted above, the orderliness of the serial hormone release process over time is typically deranged, as we have shown recently using approximate entropy (ApEn) for GH-, prolactin-, ACTH-, and aldosterone-secreting endocrine

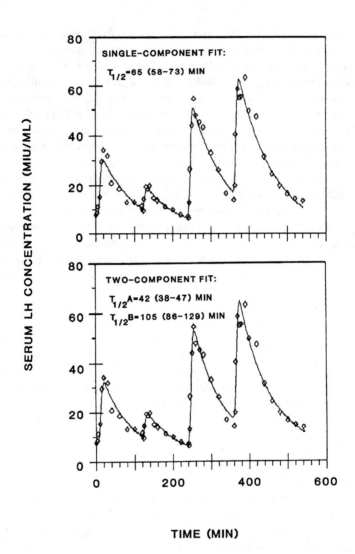

FIGURE 8

Profile of repeatedly sampled serum immunoreactive LH concentrations in GnRH-deficient man evaluated for 2 hr basally followed by 4 variably dosed i.v. GnRH injections given at 2-hr intervals. Although the visually inferred half-life of LH in the multiple peaks showing different amplitudes might appear to vary, deconvolution analysis indicates that either a single or a dual component half-life will account fully for the behavior of these data given a model of discrete punctuated LH secretory bursts fit simultaneously with the unknown LH half-life values.[56] Thus, visual inspection alone, or erroneously fitting the hormone concentration decays without allowance for ongoing secretion within the peak[28] is unreliable in judging the underlying half-life.

tumors.[39,40,57] Biomathematical models for such irregular and seemingly erratic neurohormone release from tumors have not yet been developed.

    How diurnal variations are developed from ultradian rhythms has been examined only in part, as noted above in the case of the ACTH/cortisol and the LH/testosterone axis. Further studies are clearly required to understand the complete feedback relationships and ultradian control mechanisms that generate or disrupt circadian variations, as assembled primarily from day/night variations in ultradian rhythms with or without coexistent basal (regulated or unregulated) release.

The relevance of the neurosecretory pulse signal in feedback is not yet known in most circumstances. For example, although the testis secretes androgen, estrogen, and inhibin in episodic bursts,[58,59] whether such peaks (sharp temporal variations) in plasma total and/or free testosterone concentrations significantly direct GnRH, FSH, and LH secretion by way of negative feedback is not known. In particular, how the hypothalamus and the pituitary gland would respond to a continuous testosterone negative feedback input, compared to a pulsatile testosterone input, has not been rigorously evaluated in the human. In preliminary experiments, we observe that there are significant differences in blood concentrations of testosterone and in hypothalamo-pituitary responses to continuous vs. pulsatile testosterone add-back, in healthy young men abruptly deprived of testosterone via a steroidogenic-enzyme inhibitor.[50] Further studies will be required to evaluate this question for other axes as well, e.g., the ACTH-cortisol axis, which under physiologic conditions shows strong pulsatility. Data in the human do not define whether continuous vs. pulsatile cortisol negative feedback might convey distinct information to the CRH/AVP-ACTH unit.

IV. Summary

Major advances and current tissues are reviewed here in relation to the methodology for quantitating neuroendocrine activity. Such methods encompass pulsatile data, subordinate (non-pulsatile) pattern reproducibility in neurohormone measurements, synchrony of neurohormone release, and feedback notions embracing network concepts of axis function. We have also identified a variety of unresolved issues, whereby a statement of the problem is offered for future research direction. Given the literature progress, further developments in quantitative neuroendocrinology will likely help generate novel insights into the pathophysiology of time-gated neuroendocrine regulation in health and disease.

Acknowledgments

We thank Patsy Craig for her skillful preparation of the manuscript; Paula P. Azimi for the artwork; Sandra Jackson and the nursing staff at the Clinical Research Center at the University of Virginia for excellent patient care; and Mr. David Boyd for valued assistance with Clinfo. This work was supported in part by National Institutes of Health Grant RR 00847 to the Clinical Research Center of the University of Virginia, and Research Career Development Award 1 KO4 HD 00634 (JDV).

References

1. Johnson, M.L., Veldhuis, J.D. Evolution of deconvolution analysis as a hormone pulse detection method. *Meth. Neurosci.*, 28, 1, 1995.
2. Veldhuis, J.D., Johnson, M.L. Specific methodological approaches to selected contemporary issues in deconvolution analysis of pulsatile neuroendocrine data. *Meth. Neurosci.*, 28, 25, 1995.
3. Straume, M., Veldhuis, J.D., Johnson, M.L. Model-independent quantification of measurement error: empirical estimation of discrete variance function profiles based on standard curves. *Meth. Enzymol.*, 240, 121, 1994.
4. Urban, R.J., Evans, W.S., Rogol, A.D., Kaiser, D.L., Johnson, M.L., Veldhuis, J.D. Contemporary aspects of discrete peak detection algorithms: I. The paradigm of the luteinizing hormone pulse signal in men, *Endo. Rev.*, 9, 3, 1988.
5. Sollenberger, M.L., Carlson, E.C., Johnson, M.L., Veldhuis, J.D., Evans, W.S. Specific physiological regulation of LH secretory events throughout the human menstrual cycle: new insights into the pulsatile mode of gonadotropin release. *J. Neuroendocrinol.*, 2(6), 845, 1990.
6. Evans, W.S., Christiansen, E., Urban, R.J., Rogol, A.D., Johnson, M.L., Veldhuis, J.D. Contemporary aspects of discrete peak detection algorithms: II. The paradigm of the luteinizing hormone pulse signal in women, *Endo. Rev.*, 13, 81, 1992.
7. Veldhuis, J.D., Johnson, M.L. Cluster analysis: A simple, versatile and robust algorithm for endocrine pulse detection, *Am. J. Physiol.*, 250, E486, 1986.
8. Veldhuis, J.D., Johnson, M.L. A novel general biophysical model for simulating episodic endocrine gland signaling, *Am. J. Physiol.*, 255, E749, 1988.
9. Urban, R.J., Johnson, M.L., Veldhuis, J.D. Biophysical modeling of the sensitivity and positive accuracy of detecting episodic endocrine signals, *Am. J. Physiol.*, 257, E88, 1989.
10. Veldhuis, J.D., Lassiter, A.B., Johnson, M.L. Operating behavior of dual or multiple endocrine pulse generators, *Am. J. Physiol.*, 259, E351, 1990.
11. Veldhuis, J.D., Iranmanesh, A., Johnson, M.L., Lizarralde, G. Twenty-four hour rhythms in plasma concentrations of adenohypophyseal hormones are generated by distinct amplitude and/or frequency modulation of underlying pituitary secretory bursts. *J. Clin. Endocrinol. Metab.*, 71, 1616, 1990.
12. Merriam, G.R., Wachter, K.W. Algorithms for the study of episodic hormone secretion. *Am. J. Physiol.*, 243, E310, 1982.
13. Albertsson-Wikland, K., Rosberg, S., Libre, E., Lundberg, L.O., Groth, T. Growth hormone secretory rates in children as estimated by deconvolution analysis of 24-hr plasma concentration profiles. *Am. J. Physiol.*, 257, E809, 1989.
14. De Nicolao, G., Liberati, D. Linear and nonlinear techniques for the deconvolution of hormone time-series. *IEEE Trans. Biomed. Eng.*, 40, 440, 1993.
15. Henery, R.J., Turnbull, B.A., Kirkland, M., McArthur, J.W., Gilbert, I., Besser, G.M., Rees, L.H., Tunstall Pedoe, D.S. The detection of peaks in luteinizing hormone secretion. *Chronobiol. Intl.*, 6, 259, 1989.
16. Hindmarsh, P.C., Matthews, D.R., Brain, C., Pringle, P.J., Brook, C.G. The application of deconvolution analysis to elucidate the pulsatile nature of growth hormone secretion using a variable half-life of growth hormone. *Clin. Endocrinol.*, 32, 739, 1990.

17. Jansson, P.A. Deconvolution, with Applications in Spectroscopy, *Deconvolution Methods in Spectrometry,* Academic, New York, 99, 1984.
18. Munson, P.J., Rodbard, D. *Proceedings of the Statistical Computing Section of the American Statistical Association.* Washington, D.C. 1989, 295.
19. O'Sullivan, F., O'Sullivan, J. Deconvolution of episodic hormone data: an analysis of the role of season on the onset of puberty in cows. *Biometrics,* 44, 339, 1988.
20. Pilo, A., Ferrannini, E., Navalesi, R. Measurement of glucose-induced insulin delivery rate in man by deconvolution analysis. *Am. J. Physiol.,* 233, E500, 1977.
21. Polonsky, K.S., Given, B.D., Pugh, W., et al. Calculation of the systemic delivery rate of insulin in normal man. *J. Clin. Endocrinol. Metab.,* 63, 113, 1985.
22. Prank, K., Brabant, G. Estimating thyrotropin secretory activity by a deconvolution procedure, *Meth. Neurosci.,* 20, 377, 1994.
23. Rebar, R., Perlman, D., Naftolin, F., Yen, S.S. The estimation of pituitary luteinizing hormone secretion. *J. Clin. Endocrinol. Metab.,* 37, 917, 1973.
24. Swartz, C.M., Wahby, V.S., Vacha, R. Characterization of the pituitary response in the TRH test by kinetic modeling. *Acta Endocrinol.,* 112, 43, 1986.
25. Toutain, P.L., Laurentie, M., Autefage, A., Alvinerie, M. Hydrocortisone secretion: production rate and pulse characterization by numerical deconvolution. *Am. J. Physiol.,* 255, E688, 1988.
26. Turner, R.C., Grayburn, J.A., Newman, G.B., Nabarro, J.D. Measurement of the insulin delivery rate in man. *J. Clin. Endocrinol.,* 33, 279, 1972.
27. Van Cauter, E. Method for characterization of 24-h temporal variation of blood components. *Am. J. Physiol.,* 237, E255, 1979.
28. Veldhuis, J.D., Carlson, M.L., Johnson, M.L. The pituitary gland secretes in bursts: appraising the nature of glandular secretory impulses by simultaneous multiple-parameter deconvolution of plasma hormone concentrations, *Proc. Natl. Acad. Sci. USA,* 84, 7686, 1987.
29. Veldhuis, J.D., Johnson, M.L. Deconvolution analysis of hormone data, *Meth. Enzymol.,* 210, 539, 1992.
30. Oerter, K.E., Guardabasso, V., Rodbard, D. Detection and characterization of peaks and estimation of instantaneous secretory rate for episodic pulsatile hormone secretion. *Comp. Biomed. Res.,* 19, 170, 1986.
31. Veldhuis, J.D., Moorman, J., Johnson, M.L. Deconvolution analysis of neuroendocrine data: waveform-specific and waveform-independent methods and applications. *Meth. Neurosci.,* 20, 279, 1994.
32. Veldhuis, J.D., Iranmanesh, A., Johnson, M.L., Lizarralde, G. Amplitude, but not frequency, modulation of ACTH secretory bursts gives rise to the nyctohemeral rhythm of the corticotropic axis in man. *J. Clin. Endocrinol. Metab.,* 71, 452, 1990.
33. Veldhuis, J.D., Iranmanesh, A., Lizarralde, G., Johnson, M.L. Amplitude modulation of a burst-like mode of cortisol secretion subserves the circadian glucocorticoid rhythm in man, *Am. J. Physiol.,* 257, E6, 1989.
34. Veldhuis, J.D., Liem, A.Y., South, S., Weltman, A., Weltman, J., Clemmons, D.A., Abbott, R., Mulligan, T., Johnson, M.L., Pincus, S., et al. Differential impact of age, sex-steroid hormones, and obesity on basal vs. pulsatile growth hormone secretion in men as assessed in an ultrasensitive chemiluminescence assay. *J. Clin. Endocrinol. Metab.,* 80, 3209, 1995.

35. Veldhuis, J.D., Iranmanesh, A., Clarke, I., Kaiser, D.L., Johnson, M.L. Random and non-random coincidence between luteinizing hormone peaks and follicle-stimulating hormone, alpha subunit, prolactin, and gonadotropin-releasing hormone pulsations, *J. Neuroendocrinol.*, 1, 185, 1989.

36. Veldhuis, J.D., Johnson, M.L., Seneta, E. Analysis of the co-pulsatility of anterior pituitary hormones. *J. Clin. Endocrinol. Metab.*, 73, 569, 1991.

37. Polonsky, K.S., Given, B.D., Van Cauter, E. Twenty-four-hour profiles and pulsatile patterns of insulin secretion in normal and obese subjects. *J. Clin. Invest.*, 81, 442, 1988.

38. Pincus, S.M., Keefe, D.L. Quantification of hormone pulsatility via an approximate entropy algorithm. *Am. J. Physiol.*, 262, E741, 1992.

39. Hartman, M.L., Pincus, S.M., Johnson, M.L., Matthews, D.H., Faunt, L.M., Vance, M.L., Thorner, M.O., Veldhuis, J.D. Enhanced basal and disorderly growth hormone (GH) secretion distinguish acromegalic from normal pulsatile GH release. *J. Clin. Invest.*, 94, 1277, 1994.

40. Van den Berghe, G., Pincus, S., Veldhuis, J.D., Frolich, M., Roelfsema, F. Greater disorderliness of adrenocorticotropin and cortisol release accompanies pituitary-dependent Cushing's Disease. *Eur. J. Endocrinol.*, in press, 1997.

41. Pincus, S.M., Gevers, E., Robinson, I.C.A., Roelfsema, F., Hartman, M.L., Veldhuis, J.D. Females secrete growth hormone with more process irregularity than males in both human and rat. *Am. J. Physiol.*, 270, E107, 1996.

42. Pincus, S.M., Mulligan, T., Iranmanesh, A., Gheorghiu, S., Godschalk, M., Veldhuis, J.D. Older males secrete luteinizing hormone and testosterone more irregularly, and jointly more asynchronously, than younger males: dual novel facets. *Proc. Natl. Acad. Sci. USA*, 93, 14100, 1996.

43. Veldhuis, J.D. Assessing temporal coupling between two, or among three or more, neuroendocrine pulse trains: cross-correlation analysis, simulation methods, and conditional probability testing. *Meth. Neurosci.*, 20, 336, 1994.

44. Partsch, C.-J., Abrahams, S., Herholz, N., Peter, M., Veldhuis, J.D., Sippell, W.G. Variability of pulsatile LH secretion in young male volunteers. *Eur. J. Endocrinol.*, 131, 263, 1994.

45. Martha, P.M., Rogol, A.D., Veldhuis, J.D., Blizzard, R.M. A longitudinal assessment of hormonal and physical alterations during normal puberty in boys. II. The neuroendocrine growth hormone axis during late prepuberty. *J. Clin. Endocrinol. Metab.*, 81, 4068, 1996.

46. Friend, K.E., Iranmanesh, A., Veldhuis, J.D. The orderliness of the GH release process and the mass of GH secreted per burst are highly conserved in individual men on successive days. *J. Clin. Endocrinol. Metab.*, 81, 3746, 1996.

47. Mulligan, T., Delemarre-van de Waal, H.A., Johnson, M.L., Veldhuis, J.D. Validation of deconvolution analysis of LH secretion and half-life. *Am. J. Physiol.*, 267, R202, 1994.

48. Mulligan, T., Johnson, M.L., Veldhuis, J.D. Methods for validating deconvolution analysis of pulsatile hormone release: luteinizing hormone as a paradigm. *Meth. Neurosci.*, 28, 109, 1995.

49. Veldhuis, J.D., Evans, W.S., Johnson, M.L. Complicating effects of highly correlated model variables on nonlinear least-squares estimates of unique parameter values and their statistical confidence intervals: estimating basal secretion and neurohormone half-life by deconvolution analysis. *Meth. Neurosci.*, 28, 130, 1995.

50. Zwart, A., Iranmanesh, A., Veldhuis, J.D. Disparate serum free testosterone concentrations and degrees of hypothalamo-pituitary-LH suppression are achieved by continuous vs. pulsatile intravenous androgen replacement in men: a clinical experimental model of ketoconazole-induced reversible hypoandrogenemia with controlled testosterone add-back. *J. Clin. Endocrinol. Metab.,* in press, 1997.

51. Veldhuis, J.D., Faunt, L.M., Johnson, M.L. Analysis of nonequilibrium dynamics of bound, free, and total plasma ligand concentrations over time following nonlinear secretory inputs: evaluation of the kinetics of two or more hormones pulsed into compartments containing multiple variable-affinity binding proteins. *Meth. Enzymol.,* 240, 349, 1994.

52. Veldhuis, J.D., Johnson, M.L., Faunt, L.M., Mercado, M., Baumann, G. Influence of the high-affinity growth hormone (GH)-binding protein on plasma profiles of free and bound GH and on the apparent half-life of GH. *J. Clin. Invest.,* 91, 629, 1993.

53. Hartman, M.L., Faria, A.C., Vance, M.L., Johnson, M.L., Thorner, M.O., Veldhuis, J.D. Temporal structure of *in vivo* growth hormone secretory events in man, *Am. J. Physiol.,* 260(1), E101, 1991.

54. Schaefer, F., Baumann, G., Faunt, L.M., Haffner, D., Johnson, M.L., Mercado, M., Ritz, E., Mehls, O., Veldhuis, J.D. Multifactorial control of the elimination kinetics of unbound (free) GH in the human: regulation by age, adiposity, renal function, and steady-state concentrations of GH in plasma. *J. Clin. Endocrinol. Metab.,* 81, 22, 1996.

55. Veldhuis, J.D., St.L.O'Dea, L., Johnson, M.L. The nature of the gonadotropin-releasing hormone stimulus-luteinizing hormone secretory response of human gonadotrophs *in vivo, J. Clin. Endocrinol. Metab.,* 68, 661, 1989.

56. Reyes Fuentes, A., Chavarria, M.E., Carrera, A., Aguilera, G., Rosado, A., Samojlik, E., Iranmanaesh, A., Veldhuis, J.D. Combined alterations in pulsatile LH and FSH secretion in idiopathic oligospermic men: assessment by deconvolution analysis. *J. Clin. Endocrinol. Metab.,* 81, 524, 1996.

57. Siragy, H.M., Vieweg, W.V.R., Pincus, S., Veldhuis, J.D. Increased disorderliness and amplified basal and pulsatile aldosterone secretion in patients with primary aldosteronism. *J. Clin. Endocrinol. Metab.,* 80, 28, 1995.

58. Winters, S.J., Troen, P.E. Testosterone and estradiol are co-secreted episodically by the human testis. *J. Clin. Invest.,* 78, 870, 1986.

59. Winters, S.J. Inhibin is released together with testosterone by the human testis. *J. Clin. Endocrinol. Metab.,* 70, 548, 1990.

Chapter **12**

Neuroendocrine Challenge Tests: What Can We Learn from Them?

Philip J. Cowen

Contents

0-8493-3363-6/98/$0.00+$.50
© 1998 by CRC Press LLC

I. Introduction

For those interested in understanding the pathophysiology of psychiatric disorders, the inaccessibility of the brain to direct examination has presented a formidable challenge. Neuroendocrine challenge tests can be used for several purposes, but in this chapter I will concentrate on their utility for assessing the activity of neurotransmitter pathways implicated in psychiatric illness. Many psychotropic drugs act on these neurotransmitters, and neuroendocrine challenge tests can also be used to assess both acute and chronic effect of drug treatment on neurotransmitter function.

Most neuroendocrine challenge tests in humans have measured the secretion of anterior pituitary hormones such as prolactin (PRL), growth hormone (GH) and corticotropin (ACTH). The latter hormone is frequently measured in conjunction with cortisol (CORT).[1]

II. Methodological Issues

A. Selectivity of Drug Challenge

It is, of course, possible to carry out neuroendocrine challenge tests with drugs such as insulin which produce striking increases in levels of PRL, ACTH and GH. However, hormone responses to insulin challenge are probably mediated by multiple neurotransmitter pathways. Therefore, they are not particularly useful for testing the activity of a specific neurotransmitter.

For the latter task we need a drug that selectively activates the neurotransmitter which we wish to study. The availability of selective drugs therefore becomes a key issue for the success of neurotransmitter challenge tests. Moreover, many neurotransmitter receptors exist in multiple functional subtypes. Therefore, it is necessary to decide whether one wishes to probe a particular receptor subtype or perhaps several receptors with a less selective agent.

For example, the selective noradrenaline α_2-adrenoceptor agonist, clonidine, increases plasma GH by stimulating post-synaptic α_2-adrenoceptors in the hypothalamus.[1,2] However, a similar effect can be obtained by administration of the noradrenergic re-uptake inhibitor, desipramine.[3] Desipramine indirectly activates post-synaptic α_2-adrenoceptors, presumably by blocking the re-uptake of noradrenaline into presynaptic neurones. At the same time, unlike clonidine, desipramine also increases plasma ACTH and CORT by indirect activation of post-synaptic

TABLE 1
Common Neuroendocrine Challenges for Noradrenaline (NA) and Dopamine (DA) Pathways

Drug	Hormone response	Mechanism
clonidine	↑ GH	Post-synaptic α_2-adrenoceptor agonist.
desipramine	↑ GH, ↑ ACTH/CORT, ↑ melatonin	NA re-uptake inhibitor, indirect activation of post-synaptic α_2-adrenoceptor (GH), α_1-adrenoceptor (ACTH/CORT) and β_1-adrenoceptor (melatonin).
apomorphine	↑ GH, ↓ PRL	DA_2 receptor agonist.
haloperidol	↑ PRL	DA_2 receptor antagonist.

TABLE 2
Common Neuroendocrine Challenges for 5-HT Pathways

Drug	Hormone response	Mechanism
TRP	↑ PRL, GH	5-HT precursor, indirect activation of $5\text{-}HT_{1A}$ receptors.
5-HTP	↑ CORT	5-HT precursor, indirect activation of $5\text{-}HT_2$ receptors.
clomipramine	↑ PRL	5-HT re-uptake inhibitor.
fenfluramine	↑ PRL	5-HT releaser, indirect activation of $5\text{-}HT_2$ receptors.
buspirone gepirone ipsapirone flesinoxan	↑ GH, ACTH/CORT	$5\text{-}HT_{1A}$ receptor agonist. Stimulation of post-synaptic $5\text{-}HT_{1A}$ receptors.
mCPP	↑ PRL, ACTH/CORT	$5\text{-}HT_{2C}$ receptor agonist. Stimulation of post-synaptic $5\text{-}HT_{2C}$ receptors.
sumatriptan	↓ PRL ↑ GH	$5\text{-}HT_{1B/1D}$ receptor agonist. Stimulation of pre-synaptic $5\text{-}HT_{1B}$ receptors (PRL) and post-synaptic $5\text{-}HT_{1B/1D}$ receptors (GH).

α_1-adrenoceptors.[4] It can thus provide a simultaneous measure of the responsivity of both these post-synaptic receptors. Tables 1 and 2 show a list of reasonably selective drugs that can be used to probe neurotransmitter function in humans.

B. Tolerance of Drug Challenge

Another factor relevant to neuroendocrine challenge tests is how well a drug is tolerated. Anterior pituitary hormones can be easily influenced by subjective feelings such as nausea or dizziness and if drugs produce these effects it may be uncertain how much of the increase in hormone levels is due to their specific neurotransmitter actions of a drug and how much to these non-specific "stress" factors.

Patients with certain anxiety disorders, particularly panic disorders, can be very sensitive to adverse effects of drugs 'o which they can react with paroxysmal anxiety. This may itself alter hormone levels. For example, patients with panic disorder who received the serotonin (5-HT) releasing agent, dl-fenfluramine, had both greater anxiety responses and greater increases in PRL than healthy controls.[5] However, one cannot conclude necessarily from this that panic disorder is associated with increased brain 5-HT function. It is possible for example, that fenfluramine provoked anxiety (as may many other drugs in panic disorder) and that this facilitated the elevations in plasma PRL.

C. Pharmacokinetics

It is also preferable where possible for challenging drugs to be given intravenously. This route of administration appears more likely to produce reliable endocrine responses and also avoids problems of first-pass hepatic metabolism which can lead to greatly differing plasma levels of drugs between individuals. In general, it is prudent where possible to measure plasma levels of the challenging drug.

Another advantage of intravenous drug administration is that it may avoid the effects of other drugs upon the pharmacokinetic disposition of the challenging agent. For example, selective serotonin re-uptake inhibitors (SSRIs) such as fluoxetine and paroxetine inhibit certain cytochrome P450 enzymes and can thereby elevate blood levels of co-administered drugs.[6] One study examined the effect of repeated fluoxetine treatment on the sensitivity of brain 5-HT_{2C} receptors by measuring PRL responses to the 5-HT_{2C} receptor agonist, m-chlorophenylpiperazine (mCPP).[7] However, in patients taking fluoxetine, levels of mCPP after oral administration were much higher than before treatment. PRL responses to mCPP were higher too but interpretation of this effect was complicated by the pharmacokinetic interaction between fluoxetine and mCPP.

We carried out a similar study in subjects taking paroxetine. Following repeated paroxetine administration, plasma mCPP levels after *intravenous* administration were no different to those seen before treatment. In contrast PRL responses were significantly blunted (Figure 1). In this situation, therefore, intravenous administration of mCPP avoided pharmacokinetic interaction with SSRIs and enabled the hypothesised desensitisation of 5-HT_{2C} receptors to be demonstrated.

D. Demographic Factors

Neuroendocrine challenge studies in humans require careful methodological control. Clearly hormonal responses to drug challenge can be altered by fundamental demographic factors such as age, gender and body weight.[1] The stage of the menstrual cycle can also have profound effects; for example the PRL response to d-fenfluramine is substantially greater in the luteal than the follicular phase.[8]

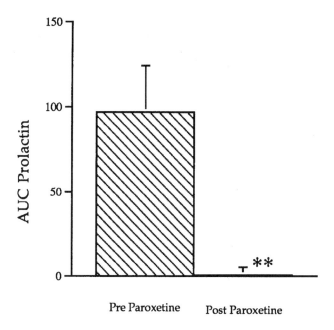

FIGURE 1

Prolactin response to mCPP (measured as area under the curve with placebo subtraction) in 7 healthy subjects tested before (pre-paroxetine) and at the end of 3 weeks paroxetine treatment (30 mg daily) (post-paroxetine). Prolactin responses post-paroxetine are significantly attenuated.

Some of these confounding factors can be minimised by the use of within-subject designs. Here, however, effects of order of testing must be considered. For example, the PRL response to dl-fenfluramine, is systematically less when subjects are retested after an interval of less than 2 weeks.[9]

E. Seasonal and Circadian Factors

Seasonal factors also need to be considered. For example, in depressed patients PRL responses to L-tryptophan (TRP) show a winter peak and a summer trough.[10] Clearly comparisons with control subjects and other groups of psychiatric patients have to be made with this in mind.

Many hormones also display a marked circadian rhythm which can alter their response to drug challenge. In neuroendocrine challenge studies, where different subjects are being compared with each other, this requires that all subjects should be tested at the same time of day. Most studies adopt a standard time of about 0900 hr after an overnight fast. However, for some hormones this may not be ideal. At this time CORT levels, for example, are usually rapidly declining and it may be relatively difficult to secure a reliable increase in CORT levels with a drug challenge.[1] In the same way PRL levels fall during the morning. PRL levels are not difficult to elevate at this time, but on the other hand it can be difficult to demonstrate a drug-induced decrease.[11]

F. Interpretation of Neuroendocrine Challenge Tests

Methodological issues are also important in interpretation of neuroendcrine challenge data. Thus a difference in neuroendocrine response to a selective drug challenge between patients and controls could reflect an abnormality in the neurotransmitter regulating the hormone concerned. It is also possible, however, that the abnormality might represent a more general change in the secretion of the hormone that is being measured. For example, the ACTH response to the $5-HT_{1A}$ receptor agonist, ipsapirone is blunted in depressed patients.[12,13] However, the ACTH response to corticotropin releasing hormone (CRH) is also blunted in depression.[14] This makes it difficult to decide whether the blunted ACTH response to ipsapirone is caused changes in $5-HT_{1A}$ receptor function or is instead a secondary consequence of abnormal HPA axis activity.

Another factor relevant to studies in psychiatric disorders is the possible effect on endocrine responses of non-specific factors such as weight loss, disturbed sleep and stress. For example, weight loss by dieting produces striking effects of 5-HT-mediated endocrine responses.[15] Many patients with depressive disorders lose weight during their illness and clearly this needs to be taken into account when 5-HT neuroendocrine tests are carried out.

A particular problem in interpretation is the effect of previous psychotropic drug treatment. Psychotropic drugs can have striking effects on responses in neuroendcrine challenge tests which can easily outlast standard washout periods of 2 to 4 weeks. For example, blunted GH responses to clonidine are often found in patients with melancholic depression.[16] However, GH responses to clonidine in normal subjects may be persistently blunted by tricyclic antidepressant treatment and this can last for many weeks. Many depressed patients who take part in studies have received previous antidepressant drug treatment. This raises the possibility that the blunted GH responses to clonidine seen in depressed patients could be due in some measure to their previous tricyclic antidepressant therapy.[17]

G. Specificity of Abnormalities

If a neuroendocrine abnormality is found in a particular patient, it is naturally tempting to speculate that this may be related in some way to the pathophysiological process. However, it is also necessary to examine whether the abnormality may be a less specific reflection of the presence of psychiatric disorder. For example, the PRL response to d-fenfluramine, is reportedly increased in schizophrenia[18] and also in chronic fatigue syndrome.[19] It would be hard to imagine two syndromes with more different psychopathology which suggests that increased brain 5-HT function does not play a key role in the pathophysiology of these disorders. Deakin and Graeff[20] have suggested that brain 5-HT pathways may play a defensive role in a

TABLE 3
Other Functional End Points of
Neuroendocrine Challenge Tests

Drug	Response
clonidine	↓ BP, ↓ MHPG, ↑ Sedation.
fenfluramine	↑ Temp, ↓ Food intake.
5-HT$_{1A}$ receptor agonists	↓ Temp.
mCPP	↑ Anxiety and autonomic activity.
	↑ Temp, ↓ Food intake.
sumatriptan	↓ Food intake, ↑ BP.

variety of aversive situations. Perhaps the presence of increased brain 5-HT in some psychiatric disorders may represent an attempt at a neurobiological level to adapt and cope. Changes in the function of particular neurotransmitter may not, therefore, be linked to specific psychiatric syndrome.

H. Other Functional End Points in Neuroendocrine Challenge Tests

As well as changes in plasma hormone levels, selective drug challenge can produce other functional changes which can be measured (see Table 3). For example, selective 5-HT receptor agonists can alter body temperature.[21] The value of such measurements is that they can provide information about aspects of neurotransmitter function brain regions other than those concerned in neuroendocrine regulation. They can therefore supplement the assessment of neurotransmitter function produced by hormonal measures.

III. Brain 5-HT Function in Major Depression

Investigation of 5-HT function in major depression gives an example of the way in which neuroendocrine challenge tests can be used to provide data on neurotransmitter function in the brain and also of some of the problems that may arise in interpretation.

It has been long been proposed that depressive disorders may be caused by a decrease in 5-HT neurotransmission.[22] This hypothesis has been based in large measure on the ability of antidepressant drugs to increase brain 5-HT function.[23] However, obtaining direct evidence that brain 5-HT function is lowered in depressed patients has proved very difficult. For example, post-mortem studies of brain 5-HT metabolism are problematic to carry out and subject to many methodological difficulties. Peripheral tissues such as plasma and blood cells can be used to investigate certain aspects of 5-HT metabolism and receptor sensitivity. However, the relevance of these peripheral models to 5-HT function in the central nervous system is uncertain (see Reference 24 for a review).

For these reasons there has been much interest in the application of 5-HT neuroendocrine tests to the study of brain 5-HT function in depression. Used in the correct way these tests can provide valuable information about 5-HT function in the living brain. As the work discussed below illustrates, neuroendocrine tests have provided the most consistent body of evidence to date that major depression is indeed associated with impaired 5-HT neurotransmission.[25]

A. Endocrine Responses to Intravenous Tryptophan

Intravenous TRP in doses of 5g or greater, reliably increases plasma concentrations of PRL and GH in humans. We have found that the PRL and GH responses to TRP are attenuated by pindolol, a β-adrenoceptor antagonist which also possesses 5-HT$_{1A}$ receptor antagonist properties. However, the endocrine effects of TRP are not inhibited by pre-treatment with selective 5-HT$_2$ or 5-HT$_3$ receptor antagonists (see Reference 21). At present, therefore, we have provisionally concluded that the PRL and GH response to TRP are mediated via indirect activation of post-synaptic 5-HT$_{1A}$ receptors, although other 5-HT receptor subtypes will need to be examined when selective probes become available.

Five studies of drug-free depressed patients have reported blunted PRL and GH responses to intravenous TRP compared to healthy controls.[25] In one study, however, the difference in PRL response in depressives could be accounted for by decreased TRP levels following infusion,[26] while in two others, diminished PRL responses could only be demonstrated when patients with recent acute weight loss were excluded.[27,28] Such exclusions are reasonable because in healthy volunteers who lose weight by dieting, there is an increase in the PRL response to TRP.[15] Accordingly, concomitant weight loss in depressed patients may obscure blunted PRL responses to TRP.

The blunted endocrine responses to TRP seen in major depression are not apparent in anxiety disorders such as panic disorder and obsessive compulsive disorder.[25] Furthermore, in bulimia nervosa blunted PRL responses to TRP in found in patients with concomitant major depression but not in those without.[29] These data suggest that impaired endocrine responses to TRP may have some specificity for depressive disorders.

B. Endocrine Responses to Clomipramine in Major Depression

Clomipramine is a selective 5-HT re-uptake inhibitor but is metabolised *in vivo* to desmethylclomipramine, a noradrenaline re-uptake inhibitor. However, during the time period of an acute clomipramine neuroendocrine challenge, plasma levels of desmethylclomipramine are not detectable.[30] Three studies in depressed patients have found that the PRL responses to clomipramine are blunted, compared to the responses of healthy controls.[25]

These data support the TRP data in suggesting that major depression is associated with impaired 5-HT-mediated PRL release. Importantly, PRL responses to other pharmacological challenges such as the dopamine D_2 receptor antagonist, metoclopramide[31] and the direct lactotroph stimulant, thyrotropin releasing hormone[30] are not reliably blunted in depressed patients. This suggests that the impairment in 5-HT-mediated PRL release seen in depression is not attributable to a generalised decrease in PRL secretion.

C. Endocrine Responses to Fenfluramine

Fenfluramine is a 5-HT releasing and uptake inhibiting agent, which has been used both as the racemate, dl-fenfluramine and the more selective 5-HT releaser, d-fenfluramine.[32] In both animals and humans, the PRL response to d-fenfluramine is abolished by the 5-HT_2 receptor antagonist, ritanserin, but not by pindolol.[33,34] This suggests that fenfluramine-induced PRL release is mediated by indirect stimulation of post-synaptic 5-HT_2 receptors.

Blunted PRL responses to fenfluramine have been found in 6 of 12 studies of drug-free depressed patients,[35] so this abnormality does not seem to be as consistent as the finding of impaired PRL response to TRP. This may because fenfluramine acts indirectly to stimulate 5-HT_2 rather than 5-HT_{1A} receptors. It also appears that blunted PRL responses to fenfluramine are associated with particular patient characteristics although studies implicate two rather different patient populations. The first tend to be inpatients, have severe depressive symptoms with melancholia and may demonstrate CORT hypersecretion.[36,37] The second have aggressive and impulsive personality traits with a history of suicide attempts.[38]

It is worthwhile noting that subjects in the latter group may manifest blunted PRL responses to fenfluramine in the absence of a current depressive disorder.[38] Presumably, here the blunted PRL responses may represent a trait marker of 5-HT dysfunction and could, perhaps, correspond with the abnormalities in CSF 5-HIAA that have been reported in subjects who tend to behave in an aggressive and impulsive way.[39] In the first group, however, where blunted PRL responses to fenfluramine are associated with severe depression and CORT hypersecretion, the impaired endocrine response appears to be a state marker of depression which remits with clinical recovery.[40]

D. Endocrine Responses to 5-HT_{1A} Receptor Agonists

The investigations outlined demonstrate that endocrine responses to drugs that activate pre-synaptic 5-HT neurones are likely to be impaired in major depression. This could reflect impaired 5-HT synthesis and release or lowered sensitivity of post-synaptic 5-HT receptors. The use of 5-HT neuroendocrine probes that directly activate post-synaptic 5-HT receptors allows this question to be addressed.

Administration of 5-HT$_{1A}$ receptor agonists to humans produces a characteristic profile of endocrine and temperature effects. The most reliable changes are an increase in plasma GH and a decrease in body temperature.[21] Given in sufficient doses, most 5-HT$_{1A}$ receptor agonists also increase plasma ACTH and CORT. Studies with pindolol suggest that the ACTH, GH and hypothermic responses to 5-HT$_{1A}$ receptor agonists are mediated by activation of 5-HT$_{1A}$ receptors.[12,21,41] Some currently employed 5-HT$_{1A}$ receptor agonists, notably buspirone and flesinoxan, also increase plasma PRL levels. However, the role of 5-HT$_{1A}$ receptors in buspirone-induced PRL release is unclear, and blockade of dopamine D$_2$ receptors may be the more important mechanism.[42]

Studies in animals suggest that endocrine responses to 5-HT$_{1A}$ receptor challenge are mediated by activation of post-synaptic 5-HT$_{1A}$ receptors.[21] The hypothermic response appears to be a consequence of activation of cell body 5-HT$_{1A}$ autoreceptors, but in some species post-synaptic 5HT$_{1A}$ receptors may be involved as well.[43]

Results of studies with 5-HT$_{1A}$ receptor agonists in drug-free depressed patients have yielded inconsistent findings. Two studies have found blunted hypothermic responses which could be consistent with lowered sensitivity of 5-HT$_{1A}$ autoreceptors.[12,44] In addition, both the studies that employed ipsapirone as a challenge found impaired CORT responses.[12,13] While this is consistent with subsensitivity of post-synaptic 5-HT$_{1A}$ receptors, depressed patients may exhibit blunted CORT responses to a variety of pharmacological and hormonal challenges, presumably because of underlying hypothalamic-pituitary-adrenal axis dysfunction (see above).

In contrast to the blunting of ipsapirone-induced CORT release, the GH response to buspirone, another probable measure of post-synaptic 5-HT$_{1A}$ receptor sensitivity, was unchanged in depressed patients.[44] This is of interest because buspirone-induced GH release is a consequence of direct activation of post-synaptic 5-HT$_{1A}$ receptors, while the GH response to TRP, which is reliably blunted in depressed patients (see above), involves indirect activation of these receptors (via increased 5-HT release). Taken together, the data suggest that the impairment in TRP-induced GH release in depression is due to abnormal function of presynaptic 5-HT neurones and not to impaired sensitivity of post-synaptic 5-HT$_{1A}$ receptors.

E. Endocrine Responses to Other 5-HT Receptor Challenges

Few studies have been reported of other 5-HT receptor challenges in depressed patients. The endocrine and temperature responses to m-chlorophenylpiperazine (mCPP) and 6-chloro-2-(1-piperazinyl)-pyrazine (MK-212) probably involve activation of 5-HT$_2$ receptors.[21] At present, there is no consistent evidence that depressed patients have altered endocrine responses to either of these drugs.[25,45]

F. Conclusions

The studies outlined above demonstrate that the use of neuroendocrine tests provides good evidence that certain aspects of brain 5-HT function are impaired in major depression. At present the most consistent abnormalities are found with drugs that act pre-synaptically to increase brain 5-HT function, particularly TRP and clomipramine. However, careful control is needed for factors such as weight loss and drug disposition. It is still uncertain whether major depression is associated with changes in the sensitivity of specific pre- and post-synaptic 5-HT receptor subtypes. However, the blunted hypothermic responses to 5-HT_{1A} receptor agonists suggest a deficit at 5-HT_{1A} raphe autoreceptors in depression. This illustrates the potential value of measuring additional functional end points during neuroendocrine challenge tests. The growing availability of selective ligands for these receptors should enable this issue to be resolved.

IV. Brain 5-HT Function During SSRI Treatment

Another use of neuroendocrine tests is to assess the effect of psychotropic drugs on brain neurotransmitter function. Here there is often animal experimental data to guide human investigations. Obviously, neuroendocrine studies in animals can be particularly useful when extrapolating analogous studies to humans.

SSRIs are increasingly used in the treatment of major depression. The acute pharmacological effects of SSRIs are essentially confined to blockade of 5-HT reuptake. This action occurs within hours of SSRI administration; however, the therapeutic effect of SSRIs can take several weeks to become manifest.[46] From this it has been argued that repeated SSRI treatment results in neuroadaptive changes in 5-HT and other receptors and it is these adaptive changes that produce the antidepressant effect.[47]

A. Animal Experimental Studies

This hypothesis has led to numerous animal experimental studies which have examined the effect of repeated SSRI treatment on pre- and post-synaptic 5-HT receptor sensitivity. There is evidence from electrophysiological studies that SSRIs decrease the sensitivity of 5-HT_{1A} autoreceptors. This would be expected to free 5-HT neurones from inhibitory feed-back control, thereby facilitating 5-HT neurotransmission.[47] The effects of SSRI treatment on post-synaptic 5-HT_{1A} receptor function in animal studies are rather contradictory and depend on the model employed. However, both neuroendocrine and behavioural studies suggest that SSRIs also decrease the sensitivity of post-synaptic 5-HT_{1A} receptors.[48,49]

There is also evidence from behavioural studies that SSRI treatment may decrease the sensitivity of other post-synaptic receptors such as 5-HT_{2A} receptors and 5-HT_{2C} receptors.[50,51] In addition, there are more limited data to suggest that SSRIs may desensitise 5-HT_{1B} terminal autoreceptors.[52] Like the desensitisation of 5-HT_{1A} autoreceptors, this action would be expected to facilitate pre-synaptic 5-HT release.[47]

From this it can be seen that the net effect of SSRIs on 5-HT neurotransmission is likely to involve a balance between a series of neuroadaptative changes in both

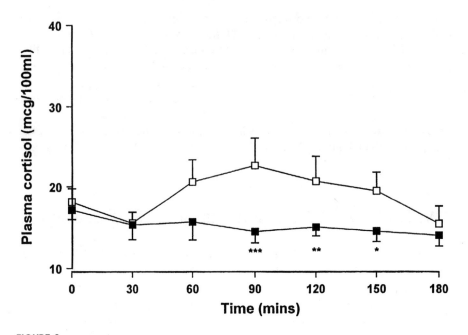

FIGURE 2
Mean plasma cortisol concentration following administration of gepirone (20 mg orally at time '0') in 12 healthy subjects who were studied before (open squares) and at the end of 3 weeks paroxetine treatment (30 mg daily) (filled squares). Cortisol levels following gepirone are significantly lower after paroxetine treatment.

pre- and post-synaptic 5-HT receptors. The desensitisation of 5-HT_{1A} and 5-HT_{1B} receptors would be expected to increase 5-HT neurotransmission while the functional down-regulation of post-synaptic 5-HT_{1A} and 5-HT_2 receptors would produce the opposite effect. As described below, it is possible to use neuroendocrine tests to assess whether SSRI treatment does alter 5-HT receptor sensitivity in humans. In addition by using 5-HT precursor challenge it is possible to assess the net effect of SSRI treatment on overall 5-HT synaptic neurotransmission.

B. SSRI Treatment and 5-HT Receptors

There is good evidence from studies of patients and healthy subjects that repeated treatment with SSRIs decreases the CORT and GH responses to 5-HT_{1A} receptor agonists such as ipsapirone, buspirone and gepirone[53,54] (Figure 2). These studies also indicate that SSRI treatment blunts the hypothermic response to 5-HT_{1A} receptor stimulation (Figure 3). Because this response appears, in part, to be mediated by pre-synaptic 5-HT_{1A} autoreceptors, this suggests that SSRI treatment in humans may desensitise 5-HT_{1A} autoreceptors. Thus these data indicate that SSRI treatment decreases the responsiveness of both pre- and post-synaptic 5-HT_{1A} receptors.

There is less work on the effect of SSRIs on other 5-HT receptor subtypes in humans. We found that the decrease in plasma PRL produced by the $5\text{-HT}_{1B/1D}$

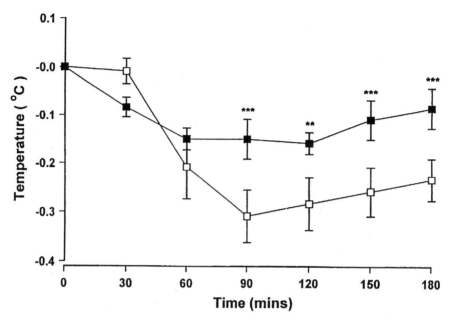

FIGURE 3

Mean oral temperature (measured as change from baseline) following administration of gepirone (20 mg orally at time '0') in 12 healthy subjects who were studied before (open squares) and at the end of 3 weeks paroxetine treatment (30 mg daily) (filled squares). Hypothermic responses following gepirone are significantly lower after paroxetine treatment.

receptor agonist, sumatriptan, was not altered by repeated SSRI treatment in healthy subjects.[55] Because sumatriptan-induced decreases in plasma PRL may be mediated by 5-HT$_{1B}$ terminal autoreceptors,[55] this suggests that SSRIs may not desensitise pre-synaptic 5-HT$_{1B}$ receptors in humans.

There have also been some neuroendocrine challenge studies on the effects of SSRI treatment on 5-HT$_{2C}$ receptor sensitivity, using the 5-HT$_{2C}$ receptor agonist, mCPP, as a pharmacological probe.[7] As explained above, the use of intravenous mCPP permitted the demonstration of blunted PRL responses to mCPP following SSRI treatment in healthy subjects (Figure 1). This suggests that in humans, as in animals, SSRI treatment lowers the sensitivity of post-synaptic 5-HT$_{2C}$ receptors.

C. SSRI Treatment and 5-HT Precursors

As noted above the endocrine responses to 5-HT precursors depends both on the pre-synaptic release of 5-HT and the sensitivity of post-synaptic 5-HT receptors. They may therefore provide an index of overall activity of 5-HT neurotransmission. In fact, studies in depressed patients with both TRP and the 5-HT precursor, 5-HTP, show that repeated SSRI treatment increases 5-HT-mediated endocrine responses to 5-HT precursors[56] (Figure 4).

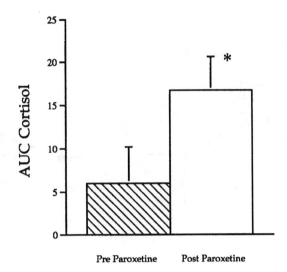

FIGURE 4

Cortisol response to 5-HTP (measured as area under the curve with placebo subtraction) in 10 depressed patients tested before (pre-paroxetine) and at the of 8 weeks paroxetine treatment. Cortisol responses post-paroxetine are significantly enhanced.

D. Conclusions

Neuroendocrine studies in humans indicate that SSRIs do indeed produce adaptive changes in certain pre- and post-synaptic 5-HT receptors. While the net effects of the various changes on 5-HT neurotransmission is hard to predict, studies with 5-HT precursors suggest that overall SSRIs increase brain 5-HT function. This is consistent with the ability of techniques that lower brain 5-HT function, e.g., the process of TRP depletion, to produce a rapid reversal of the antidepressant effects of SSRIs in depressed patients.[57]

V. Utility of Neuroendocrine Challenge Tests in Humans

Direct measurement of neurotransmitter activity in the human brain will continue to pose methodological problems. Advances in brain imaging will go some way to alleviating this situation. For example, the development of selective ligands in conjunction with Single Photon Emission Tomography and Positron Emission Tomography will allow clear delineation of various neurotransmitter receptors and their distinct subtypes.[58] However, this will not necessarily provide information about receptor function, which is the critical factor in determining synaptic neurotransmission.

Brain imaging can also be used in conjunction with pharmacological probes to investigate neurotransmitter function. Even here, however, the functional end-point (for example, a change in regional blood flow blood-flow) is somewhat indirect and remote from the site of pharmacological action.[59] Accordingly, it is liable to difficulties in interpretation. For this reason neuroendocrine challenge tests are likely to retain a place as a relatively simple and acceptable means of measuring neurotransmitter function in the human brain.

Acknowledgment

The author is a Medical Research Council Clinical Scientist.

References

1. Checkley, S. A. Neuroendocrine tests of monoamine function in man: a review of basic theory and its application to the study of depressive illness, *Psychol. Med.,* 10, 35-53, 1980.
2. Eriksson, E., Eden, S. and Modigh, K. Up- and down-regulation of central post-synaptic a_2-adrenoceptors reflected in GH response to clonidine in the reserpine treated rat, *Psychopharmacology,* 77, 327, 1982.
3. Laakmann, G., Zygan, K., Schon, H. W., Weiss, A., Wittmann, M., Meissner, R. and Blaschke, D. Effects of receptor blockers (methysergide, propranolol, phentolamine, yohimbine and prazosin) on desipramine-induced pituitary hormone stimulation in humans. 1: Growth hormone, *Psychoneuroendocrinology,* 11, 447-461, 1986.
4. Laakmann, G., Wittmann, M., Schon, H. W., Zygan, K., Weiss, A., Meissner, R., Mueller, E. A. and Stalla, G. K. Effects of receptor blockers (methysergide, propranolol, phentolamine, yohimbine and prazosin) on desipramine-induced pituitary hormone stimulation in humans. 3: Hypothalamo-pituitary-adrenocortical axis, *Psychoneuroendocrinology,* 11, 475-489, 1986.
5. Targum, S. D. Differential responses to anxiolgenic challenge studies in patients with major depressive disorder and panic disorder, *Biological Psychiatry,* 28, 21-34, 1990.
6. Nemeroff, C. B., De Vane, L. and Pollock, B. G. Newer antidepressants and cytochrome P450 system, *Am. J. Psychiatry,* 153, 311-320, 1996.
7. Hollander, E., de Caria, C., Gully, R., Nitescu, A., Suckow, R. F., Gorman, J. M., Klein, D. F. and Liebowitz, M. R. Effects of chronic fluoxetine treatment on behavioural and neuroendocrine responses to meta-chlorophenylpiperazine in obsessive compulsive disorder, *Psychiatry Res.,* 36, 1-17, 1991.
8. O'Keane, V., O'Hanlon, M., Webb, M. and Dinan, T. D-fenfluramine/prolactin response throughout the menstrual cycle: evidence for an oestrogen induced alteration, *Clin. Endocrinol.,* 34, 289-292, 1991.
9. Coccaro, E. F., Siever, L. J., Klar, H., Rubenstein, K., Benjamin, E. and Davis, K. L. Diminished prolactin responses to repeat fenfluramine challenge in man, *Psychiatry Res.,* 22, 257-259, 1987.

10. Cappiello, A., Malison, R. T., McDougle, C. J., Vegso, S. J., Charney, D. S., Heninger, G. R. and Price, L. H. Seasonal variation in neuroendocrine and mood responses to IV l-tryptophan in depressed patients and healthy subjects, *Neuropsychopharmacology*, 15, 475-483, 1996.

11. Boeles, S., Williams, C., Campling, G. M., Goodall, E. M. and Cowen, P. J. Sumatriptan decreases food intake and increases plasma growth hormone in healthy women, *Psychopharmacology*, 129, 179-182, 1997.

12. Lesch, K. P. 5-HT$_{1A}$ receptor responsivity in anxiety disorders and depression, *Prog. Neuropsychopharmacol. Biol. Psychiatry*, 15, 723-733, 1992.

13. Meltzer, H. Y. and Maes, M. Effect of ipsapirone on plasma cortisol and body temperature in major depression, *Biol. Psychiatry*, 38, 450-457, 1995.

14. Gold, P. W., Chrousos, G., Kellner, C., Post, R., Roy, A., Augerinos, P., Schulte, H., Oldfield, E. and Loriaux, D. L. Psychiatric implications of basic and clinical studies with corticotropin releasing factor, *Am. J. Psychiatry*, 5, 619-627, 1984.

15. Anderson, I. M., Parry-Billings, M., Newsholme, E. A., Fairburn, C. G. and Cowen, P. J. Dieting reduces plasma tryptophan and alters brain 5-HT function in women, *Psychol. Med.*, 20, 785-791, 1990.

16. Checkley, S. A., Glass, I. B., Thompson, C., Corn, T. and Robinson, P. The GH response to clonidine in endogenous as compared with reactive depression, *Psychol. Med.*, 14, 773-777, 1984.

17. Schittecatte, M., Charles, G., Machowski, R. and Wilmott, E. Tricyclic washout and growth hormone response to clonidine, *Brit. J. Psychiatry*, 154, 853-863, 1989.

18. Abel, K. M., O'Keane, V. and Murray, R. M. Enhancement of the prolactin response to d-fenfluramine in drug-naive schizophrenic patients, *Brit. J. Psychiatry*, 168, 57-60, 1996.

19. Cleare, A. J., Bearn, J., Allain, T., McGregor, A., Wessely, S., Murray, R. M. and O'Keane, V. O. Contrasting neuroendocrine responses in depression and chronic fatigue syndrome, *J. Affect. Disorders*, 35, 283-289, 1995.

20. Deakin, J. F. W. and Graeff, F. G. 5-HT and mechanisms of defence, *J. Psychopharmacol.*, 5, 305-315, 1991.

21. Cowen, P. J. Serotonin receptor subtypes in depression: evidence from studies in neuroendocrine regulation, *Clin. Neuropharmacol.*, 16(Suppl 3), S6-S18, 1993.

22. Coppen, A. J. The biochemistry of affective disorders, *Brit. J. Psychiatry*, 113, 1237-1264, 1967.

23. Cowen, P. J. A role for 5-HT in the action of antidepressant drugs, *Pharmacol. Therapeut.*, 46, 43-51, 1990.

24. Cowen, P. J. The serotonin hypothesis: necessary but not sufficient, in *Selective Serotonin Re-uptake Inhibitors. 2nd ed.*, Feighner, J. P. and Boyer W. F., Eds., John Wiley, Chichester, 1996, 63-86.

25. Power, A. C. and Cowen, P. J. Neuroendocrine challenge tests: assessment of 5-HT function in anxiety and depression, *Mol. Aspects Med.*, 13, 205-220, 1992.

26. Koyama, T. and Meltzer, H. Y. A biochemical and neuroendocrine study of the serotonergic system in depression, in *New Results in Depression*, Hippius, H., Klerman, G. L. and Matussek, N., Eds., Springer, Berlin, 1986:169-188.

27. Cowen, P. J. and Charig, E. M. Neuroendocrine responses to intravenous tryptophan in major depression, *Arch. Gen. Psychiatry,* 44, 958-966, 1987.

28. Deakin, J. F. W., Pennell, I., Upadhyaya, A. J. and Lofthouse, R. A neuroendocrine study of 5-HT function in depression: evidence for biological mechanisms of endogenous and psychosocial causation., *Psychopharmacology,* 101, 85-92, 1990.

29. Brewerton, T. D., Mueller, E. A., Lesem, M. D., Brandt, H. A., Quearry, B., George, T., Murphy, D. L. and Jimerson, D. C. Neuroendocrine responses to *m*-chlorophenylpiperazine and l-tryptophan in bulimia, *Arch. Gen. Psychiatry,* 49, 852-861, 1992.

30. Anderson, I. M., Ware, C. J., da Roza Davis, J. M. and Cowen, P. J. Decreased 5-HT-mediated prolactin release in major depression, *Brit. J. Psychiatry,* 160, 372-378, 1992.

31. Anderson, I. M. and Cowen, P. J. Prolactin response to the dopamine antagonist, metoclopramide, in depression, *Biol. Psychiatry,* 30, 313-316, 1991.

32. McTavish, D. and Heel, R. C. Dexfenfuramine, *Drugs,* 43, 713-733, 1992.

33. Goodall, E. M., Cowen, P. J., Franklin, M. and Silverstone, T. Ritanserin attenuates anorectic, endocrine and thermic responses to d-fenfluramine in human volunteers., *Psychopharmacology,* 112, 461-466, 1993.

34. Park, S. B. G. and Cowen, P. J. Effect of pindolol on the prolactin response to d-fenfluramine., *Psychopharmacology,* 118, 471-474., 1995.

35. Park, S. B. G., Williamson, D. J. and Cowen, P. J. 5-HT neuroendocrine function in major depression: prolactin and cortisol responses to d-fenfluramine, *Psychol. Med.,* 26, 1191-1196, 1996.

36. Mitchell, P. and Smythe, G. Hormonal responses to fenfluramine in depressed and control subjects, *J. Affect. Disorders,* 19, 43-51, 1990.

37. Lichtenberg, P., Shapira, B., Gillon, D., Kindler, S., Cooper, T. B., Newman, M. E. and Lerer, B. Hormone responses to fenfluramine and placebo challenge in endogenous depression, *Psychiatry Res.,* 43, 137-146, 1992.

38. Coccaro, E. F., Siever, L. J., Klar, H. M., Maurer, G., Cochrane, K., Cooper, T. B., Mohs, R. C. and Davis, K. L. Serotonergic studies in patients with affective and personality disorders, *Arch. Gen. Psychiatry,* 46, 587-599, 1989.

39. Roy, A., Virkkunen, M. and Linnoila, M. Serotonin in suicide, violence and alcoholism, in *Serotonin in Major Psychiatric Disorders,* Coccaro, E. F. and Murphy, D. L., Eds., American Psychiatric Press, Washington, D.C., 1990, 187-208.

40. Shapira, B., Cohen, J., Newman, M. E. and Lerer, B. Prolactin response to fenfluramine and placebo challenge following maintenance pharmacotherapy withdrawal in remitted depressed patients, *Biol. Psychiatry,* 33, 531-535, 1993.

41. Seletti, B., Benkelfat, C., Blier, P., Annable, L., Gilbert, F. and de Montigny, C. Serotonin$_{1A}$ receptor activation by flesinoxan in humans: body temperature and neuroendocrine responses, *Neuropsychopharmacology,* 13, 93-104, 1995.

42. Meltzer, H. Y., Gudelsky, G. A., Lowy, M. T., Nash, J. F. and Koenig, J. I. Neuroendocrine effects of buspirone: mediation by dopaminergic and serotonergic mechanisms, in *Buspirone: Mechanisms and Clinical Aspects,* Tunnicliff, G., Eison, A. S. and Taylor, O. P., Eds., Academic Press, San Diego, 1991, 177-192.

43. Bill, D. J., Knight, M., Forster, E. A. and Fletcher, A. Direct evidence for an important species difference in the mechanism of 8-OH-DPAT-induced hypothermia, *Brit. J. Pharmacol.,* 103, 1857-1864, 1991.

44. Cowen, P. J., Power, A. C., Ware, C. J. and Anderson, I. M. 5-HT$_{1A}$ receptor sensitivity in major depression: a neuroendocrine study with buspirone, *Brit. J. Psychiatry*, 164, 372-379, 1994.

45. Anand, A., Charney, D. S., Delgado, P. L., McDougle, C. J., Heninger, G. R. and Price, L. H. Neuroendocrine and behavioral responses to intravenous m-chlorophenylpiperazine (mCPP) in depressed patients and healthy comparison subjects, *Am. J. Psychiatry*, 151, 1626-1630, 1994.

46. Asberg, M., Erikkson, B., Matensson, B., Traskman-Bendz, L. and Wagner, A. Therapeutic effects of serotonin uptake inhibitors in depression, *J. Clin. Psychiatry*, 47, 23-35, 1986.

47. Blier, P. and de Montigny, C. Current advances and trends in the treatment of depression, *Trends Pharmacol. Sci.*, 15, 220-226, 1994.

48. Goodwin, G. M., de Souza, R. J. and Green, A. R. Attenuation by electroconvulsive shock and antidepressant drugs of the 5-HT$_{1A}$ receptor mediated hypothermia and serotonin syndrome produced by 8-OH-DPAT in the rat, *Psychopharmacology*, 91, 500-505, 1987.

49. Li, Q., Brownfield, M. S., Levy, A. D., Battaglia, G., Cabrera, T. M. and Van de Kar, L. D. Attenuation of hormone responses to the 5-HT$_{1A}$ agonist, ipsapirone, by long-term treatment with fluoxetine but not desipramine in male rats, *Biol. Psychiatry*, 36, 300-308, 1994.

50. Johnson, A. M. The comparative pharmacological properties of selective serotonin re-uptake inhibitors in animals, in *Selective Serotonin Re-uptake Inhibitors*, Feighner, J. P. and Boyer, W. F., Eds., John Wiley, Chichester, 1991, 37-70.

51. Kennett, G. A., Lightowler, S., de Biasi, V., Stevens, N. C., Wood, M. D., Tulloch, I. F. and Blackburn, T. P. Effect of chronic administration of selective 5-hydroxytryptamine and noradrenaline uptake inhibitors on a putative index of 5-HT$_{2C/2B}$ receptor function, *Neuropharmacology*, 33, 1581-1588, 1994.

52. Blier, P. and Bouchard, C. Modulation of 5-HT release in the guinea-pig brain following long-term administration of antidepressant drugs, *Brit. J. Pharmacol.*, 113, 485-495, 1994.

53. Lesch, K. P., Ho, H. A., Schulte, H. M., Osterheider, M. and Muller, T. Long-term fluoxetine treatment decreases 5-HT$_{1A}$ receptor responsivity in obsessive compulsive disorder, *Psychopharmacology*, 105, 415-420, 1991.

54. Anderson, I. M., Deakin, G. F. W. and Miller, A. T. J. The effects of chronic fluvoxamine on hormonal and psychological responses to buspirone in normal volunteers, *Psychopharmacology*, 128, 74-82, 1996.

55. Wing, Y.-K., Clifford, E. M., Sheehan, B. D., Campling, G. M., Hockney, R. A. and Cowen, P. J. Paroxetine treatment and the prolactin response to sumatriptan, *Psychopharmacology*, 124, 377-379, 1996.

56. Price, L. H., Charney, D. S., Delgado, P. L., Anderson, G. M. and Heninger, G. R. Effects of desipramine and fluvoxamine treatment on the prolactin response to tryptophan: serotonergic function and the mechanism of antidepressant action, *Arch. Gen. Psychiatry*, 46, 625-631, 1989.

57. Delgado, P. L., Charney, D. S., Price, L. H., Aghajanian, G. K., Landis, H. and Heninger, G. R. Serotonin function and the mechanism of antidepressant action: reversal of antidepressant-induced remission by rapid depletion of plasma tryptophan, *Arch. Gen. Psychiatry,* 47, 411-418, 1990.

58. Sedvall, G., Farde, L., Persson, A. and Wiesel, F. A. Imaging of neurotransmitter receptors in the living human brain, *Arch. Gen. Psychiatry,* 43, 995-1005, 1986.

59. Grasby, P. M., Friston, K. J., Bench, C., Cowen, P. J., Frith, C. D., Liddle, P. F., Frackowiak, R. S. J. and Dolan, R. J. Effect of the 5-HT$_{1A}$ partial agonist buspirone on regional cerebral blood flow in man, *Psychopharmacology,* 108, 380-386, 1992.

Index

A

Acetylcholine, 71–72
ACTH. *See* Adrenocorticotropin
Actin, 25
Action potentials, 96–99, 101, 103, 105, 118
Activin, 43–46
Adenohypophysial hormone release, 97, 109, 147
Adenosine triphosphate (ATP), 19, 21, 27, 100
Adenylate cyclase, 43
Adrenal cortical hormone response, 164, 171
Adrenal corticosteroid (CORT)
    depression and, 213–214, 216
        drug challenge tests and, 206–207, 209
        stress and, 164, 168–169, 172–175
Adrenal medulla, 147
Adrenergic receptors, 73, 154
Adrenoceptors, 147, 206–207, 212
Adrenocorticotropin (ACTH)
    amine neurotransmitters and, 148–155
    antigens, 40
    biotinylated ligands and, 34, 40–42, 46
    drug challenge tests and, 206–207, 210, 214
    feedback sensitivity and, 52
    release of, 189–191, 193, 197–199
    steroid hormones and, 51, 63
    stress and, 147, 164, 166, 169–171, 174
Affinity cytochemistry, 32, 46
Afterhyperpolarization, 97
Afterpotentials, 97–99, 104
Agarose gel, 20
Age, hormone release and, 192, 208
Alarm phase, 169
Aldosterone, 197
Allopregnanolone, 51
Alpha amino carbazol, 38
Alternatively spliced transcripts, 18
Amines, neuroendocrine regulation and, 145–156
Amino acids, 72, 99
Amphibians, 151
AMV-RT (avian myeloblastosis virus-reverse
    transcriptase), 19, 21

Amygdala, 135, 174–175
Anabolic effects of stress, 171
Analgesia, stress-induced, 171
Analysis of variance, 154
Anatomical identification, 107–108, 110
Anatomical resolution, 54–56, 60, 108
Androgen, 50–51, 55
Androgen receptors, 52, 54–55, 57–58, 61–63
Androgen receptor immunoreactivity (AR-ir),
    55–56
Anesthetized recording preparations, 117–118, 121
Angiotensin converting enzyme, 170
Angiotensin II (AII), 71, 134, 139
Angiotensinogen, 137
Angiotensin receptors, 28, 133, 137–138
Animal handling, 166, 173
Annealing temperature, 23–24, 26
Anterior cerebral artery, 73
Anterior hypothalamic tumors, 3–4, 6
Anterior pituitary
    biotinylated ligands and, 32
    drug challenge tests and, 207
    explants and, 75
    feedback sensitivity and, 52
    hormone secretion and, 151
    LHRH neurons and, 12, 74
    luteinizing hormone and, 1
    RT-PCR and, 25, 27–28
    steroid hormones and, 56
    stress hormones and, 164
Anterior supraoptic nucleus (SONa), 103
Antidepressants, 210, 215
Antidromic activation, 73, 103–104, 122
Antimitotic agents, 78
Antisense oligodeoxynucleotides (ODNs),
    131–141
    administration of, 135
    controls for, 136–138
    effectiveness of, 133–135
    mechanisms of action of, 132–133, 138–141
    scramble, 136–140
    sense and, 136

225

dispersed hypothalamic cell preparations and,
   77–80
explants and, 70–72, 76
neuroendocrine cells and, 96–97, 104,
   107–109, 118, 127
pulsatile secretion of, 118
stress and, 169–170, 174

P

Pain, 156, 166, 172
Pancreatic β-cell tumors, 3
Panic disorder, 212
Paracrine interaction, 36, 151
Paraformaldehyde, 35–36, 38
Paraventricular nucleus (PVN)
   amine neurotransmitters and, 146–148, 150,
      155
   antisense treatment and, 133–134, 137–139
   explant preparations of, 72, 74, 76
   intracellular recording and, 103, 105–107
   lesions of, 148–151
   in vivo studies and, 121–122, 126
Parenchymal infusion, 135
Paroxetine, 208–209, 216–218
Parvocellular neuroendocrine cells, 97, 105,
   108–110, 118, 146
Patch clamp recording, 81, 99–102
Patch pipettes, 99–100
Pavlovian responses, 172–173
PCR (polymerase chain reaction), 17–28, 108
32P-dATP labels, 19, 27
Pentobarbital, 117–118
Peptide pool, 134
Peptide secretion, 139–140, 164
Peptide synthesis, 34, 133
Perifusion approach, 70, 72–74, 82
Perinuclear zone, 71
PG-21, 56
Pharmacokinetics, 208
Phenol-chloroform-isoamyl alcohol extraction, 19
Phosphodiester (PO) antisense, 132, 136
Phosphorimagers, 19, 27
Phosphorothioate (PS) antisense, 132, 135–136
Photolysis of caged glutamate, 105–106
Photostimulation, 105–106
Piloerection, 164
Pindolol, 212–213
Pituitary
   anterior. *See* Anterior pituitary
   antisense treatment and, 140
   biotinylated ligand identification, 31–46
   collection and plating of cells from, 33–34
   estrogen receptors in, 57
   GT-1 cells and, 12

intermediate lobe, 151
NPY Y1 receptors in the, 18, 27
posterior, 71–72
stress and, 171
Pituitary-adrenocortical axis, 167–168
Pituitary cells, 12, 34, 36–37
Pituitary stalk, 70, 103, 148
Platelet-derived growth factor, 136
Plating density, 82
Plating methods, 6, 33–34, 82
POA (preoptic area), 74
Polyacrylamide gel electrophoresis, 19, 27, 55, 57,
   60–61
Poly-lysine, 82
Polymerase chain reaction (PCR), 17–28, 108
Polyribosomes, 8
PO (phosphodiester) antisense, 132, 136
Positron Emission Tomography, 218
Post-mortem studies, 211
Postsynaptic currents, 99, 101
Post-traumatic stress disorder (PTSD), 176
Potassium chloride, 74
Prenatal animals, 74
Preoptic area (POA), 74
Preoptic-hypothalamic LHRH neurons, 9, 74
Preovulatory surge, 24
Pressure ejection techniques, 125
Primary anterior pituitary cells, 12
Primary hypothalamic cultures, 77–83
Primates, 176
Primer-dimer formation, 23
Primers for PCR, 19–24, 26
PRL. *See* Prolactin
Proestrus, 37, 40, 42–44
Progesterone, 50–51, 82, 125, 127
Progesterone receptors, 58–59, 62–63
Prolactin (PRL)
   amine neurotransmitters and, 148, 166
   in biotinylated studies, 41–42
   in depression, 212–214, 216
   drug challenge tests and, 208–209
   in explant studies, 75
   release of, 197
   stress and, 166, 170–171, 173–175
Pro-opiomelanocortin, 6, 75, 78, 147, 170
Propranolol, 170
Prostaglandins, 12, 71
Prostate gland, 53, 61
Protamine sulfate, 54
Protease inhibitors, 34
Protein A conjugated red blood cells, 36
Protein Kinase C, 43
Protein synthesis, antisense action and, 133
Proto-oncogenes, 136
Pseudogenes, 22